T0075155

Program Management

Program Management

Going Beyond Project Management to Enable Value-Driven Change

Dr. Al Zeitoun, PgMP, PMI Fellow
Global Strategy Execution Executive
Bethesda, Maryland, USA

WILEY

Published by John Wiley & Sons, Inc., Hoboken, New Jersey.
Published simultaneously in Canada.

Library of Congress Cataloging-in-Publication Data Applied for:

Hardback ISBN: 9781119931287

Cover Design: Wiley
Cover Image: © geralt/Pixabay

Set in 10.5/13pt STIXTwoText by Straive, Chennai, India

SKY10056309_092823

To

my wife, Nicola

and my kids, Adam, Zeyad, and Sarah

for being my continued source of inspiration

the memory of

my Dad and Mom

for instilling in me the value of learning

my life's mentor

Dr. Harold Kerzner

for demonstrating impactful thought leadership

Contents

Preface

For more than 50 years, project and program management have been in use, perhaps not on a worldwide basis. Significantly more literature appeared describing the accomplishments of project management rather than program management. By the end of this decade, there will be a great deal of growth in program management applications as companies begin working on more projects, especially those related to strategic business initiatives, innovation, and R&D. Grouping these projects into programs will be a necessity to maximize the expected business benefits and value.

What differentiated companies in the early years was whether they used project management, not how well they used it. Today, almost every company uses project management, and the differentiation is whether they are simply good at project management or whether they truly excel at project management. As the acceptance and use of program management grows, the difference between being good and excelling at program management is expected to be quite large. Program management is heavily focused on business applications. Project management is generally seen as a short-term perspective, but program management is often viewed in a 10–20-year time frame.

Companies such as IBM, Microsoft, Siemens, Hewlett-Packard, and Deloitte, just to name a few, have come to the realization that they must excel at project management. As such, these companies encourage their employees to become certified in project management. As program management applications increase, companies are expected to encourage their workers to become certified in program management practices as well because of the significant return on investment it can have on business success.

History is a great teacher and predictor of the future. What we learned from project management was that success was measured as a continuous stream of successfully managed projects. The challenge now will be defining program management success. Some of the components that are expected to be included in program management success are:

- *Nonlinear thinking*: Project management traditionally focuses on linear thinking with well-defined requirements and life cycle phases. Companies use the same forms, guidelines, templates, and checklists on all projects. The projects within programs may begin with just an idea, and new policies and procedures may need to be developed as programs progress. Projects within a program can become highly complex over time.
- *Strategic focus*: The focus of most projects is on the creation of a deliverable acceptable to stakeholders. The focus of programs will be on the creation of long-term business benefits and business value.
- *Change management*: Programs have a much greater impact on the organization's business models than do individual projects. As such, program managers must continuously evaluate the long-term effects of a successful program and prepare for change as early as possible. Program change management necessities have a much greater likelihood of removing people from their comfort zones.
- *Collaboration*: Traditional project may have just a few stakeholders, and their involvement in the project may be minimal. Programs may have a multitude of stakeholders, many of whom desire to be involved in critical decision-making. Programs therefore require much more collaboration and engagement, often on a continuous basis, than traditional project management.
- *Methodologies and life cycle phases*: Program managers must be able to manage multiple projects within a program, where each project can use a different methodology and have different life cycle phases.
- *Integration*: Each of the projects within a program must be integrated with other projects within the program. This creates challenges for program managers with problem-solving and decision-making.

- *Risk management*: The VUCA environment has a much greater impact on risk management and eventual success for programs than for individual projects. The program manager must continuously monitor the VUCA environment and be willing to mitigate risks, especially business-related risks that might impact future business value creation.
- *Culture*: Unlike project managers who may focus on just one project culture, program managers must deal with multiple cultures and how each culture interfaces with the corporate culture. Program managers must also possess the leadership skills to deal with cultures that can change over the long-term life cycles of programs.
- *Problem-solving and decision-making*: The growth in the number of projects will make it difficult for the program manager to solve problems and make decisions without the use of digitalization technology such as Artificial Intelligence (AI). On traditional projects, the project manager may have the luxury of deciding whether or not to use AI practices. But on programs, given the possibility of a large number of projects being performed consecutively, the program manager may find it necessary to educate all of the stakeholders on how digitalization practices will affect their projects.

There are, of course, other topics that will help in defining and achieving program management success. All of the topics discussed above, and more, appear in Dr. Zeitoun's book. If your company wishes to achieve the program management success that other companies have found, then this book is a "must read."

The future of program management may very well rest in the hands of solution providers. These providers will custom-design program management practices, such as those discussed in this book, for each client and possibly for each stakeholder. They must be able to develop program management skills that go well beyond current program management documentation and demonstrate a willingness to drive change, and make strategic business decisions as well as program decisions. The future of program management looks quite promising, but it will be challenging.

Dr. Harold Kerzner
Senior Executive Director for Project Management
International Institute for Learning, Inc. (IIL), USA
September 2023

Introduction

The Why of Program Management

BACKGROUND

The world has been changing and will continue to change and fast. If we have learnt anything from that last decade's experiences, it is that change will accelerate in its scope and speed and that we best be ready to adapt and to continually build the necessary future skills so as not only to sustain our progress but also to grow and lead. Projects have and will continue to be the vehicle for change. In its simplest form, a project is an endeavor that has a beginning and an end and consumes resources toward achieving a specific and meaningful change purpose.

Programs become the connecting grouping of projects to make the best coordinated use of precious resources across the projects and toward achieve benefits (something of value to the key program stakeholders) that would otherwise be missed if we manage these projects individually. Program stakeholders are the ones with interest in the program's outcomes and in many instances have the power to influence the program's direction toward achieving its benefits. The key word here is benefit, which ultimately becomes the key to achieving value. This is what makes programs exciting. They become the closest strategic vehicle leaders have to plan and execute large and sometimes complex change efforts. Programs are in essence the natural cascade from strategic vision into the reality of achieving what matters.

Another key aspect to the background of the importance of programs is that they represent the big nuggets in each portfolio that an organization would typically have in place to execute its strategy. A portfolio is in essence the holistic bucket of programs, components thereof, projects, and other parts of the business operations. Organizing the portfolio in a way that links directly to achieving the mission and vision of the organization becomes critical. Two building blocks exist in a given portfolio. One is focused on running the business. The other is focused on changing the business.

The main focus of this textbook is on strategic business opportunities in the form of programs and other strategic projects, changing the business. The "Why" of program management is in essence the criticality of ensuring that we step beyond just achieving deliverables, which seems to be the biggest focus of projects, to what most counts, which is getting to the change results that are envisioned in the programs' choices that are designed to achieve these benefits.

The book opens the door to what I could refer to as the ***Program Way***. It is a mindset and a way of working that is centered around finding the most fitting and simplest ways of working to handle the possible complexity with these likely major strategic business opportunities. While building on the classical project and program management tools and principles, this ***Program Way*** opens the door to multiple creative ideas for how we deliver project and program management efforts in the future. Scaling strategy execution services, for example by considering Program or Project Management Office, (PMO)-as-a-Service, will continue to be a strategic priority for organizations as they strive to achieve the most impactful mission that affect how we live, work, consume, and change societies.

Program Management: Going Beyond Project Management to Enable Value-Driven Change, First Edition. Al Zeitoun.

DIFFERENCES BETWEEN PROGRAMS AND PROJECTS

Introduction

I worked with Dr. Harold Kerzner on this introductory portion of this book as we wanted to jointly address a few of the key differences between programs and projects. From the beginning of modern project management, there has been considerable confusion concerning the relationship between projects and programs, and how they are managed. The terms have often been used interchangeably. There have been articles written on the differences (Weaver 2010). But now, partially due to the learning from the COVID-19 pandemic, organizations are looking much more closely at the differences and whether with their limited funding, accompanied by a loss of critical resources, they should focus more on program rather than project management efforts.

The PMBOK® Guide provides the following definitions (Project Management Institute 2021)

- *Project*: A temporary endeavor undertaken to create a unique product, service, or result.
- *Program*: Related projects, subsidiary programs, and program activities that are managed in a coordinated manner to obtain benefits not available from managing them individually.

From a cursory position, projects focus on the creation of unique deliverables usually for a single customer or stakeholder. Programs focus on the synergistic opportunities that can be obtained from managing multiple projects to create business benefits and business value for both the organization and its customers.

There are textbooks that simply define program management as the management of multiple projects. However, there are other factors that create significant differences between projects and programs. Managing a project as if it was a program can lead to significant cost overruns and cancellation. Managing a program as if it was a project can lead to significantly less than optimal results and failure.

There are numerous factors that can be used to differentiate programs from projects. Some commonly used factors to identify the boundaries of projects and programs include:

- Type of objectives
- Type of products and services produced
- Industry type and characteristics
- Number and types of customers/stakeholders that will benefit
- Impact on business success and definition of business success
- Strategic risks
- Methodology used for implementation
- Size of the project or program
- Impact of enterprise environment factors such as the VUCA environment
- Complexity of the requirements
- Technology required and availability
- Strategic versus operational decision-making

Perhaps the greatest difference however is in the organizational behavior factors, the leadership style selected, the interaction with the team and stakeholders, and the decisions that must be made. Several of these factors are discussed in this study.

Selecting the Leader

There is a mistaken belief among many companies that the most important criteria for becoming a project or program manager is simply becoming certified through an examination process. While certification is an important factor, there are other attributes that may be considerably more important.

A vice president in an aerospace company commented that the two most important skills for a PM in his organization were a command of technology and writing skills. The command of technology had to be specifically related to the technology required to produce the project's deliverables. The assignment as a PM could be temporary, without any training, just for this project, and the PM could then return to his/her functional organization for other duties perhaps not project management related. Project management was seen as a part-time rather than a full-time career path position. Some companies hire contractors to perform work in their organizations and the contractors then take the responsibility for project management.

Program management must be owned by the organization and usually a full-time assignment. Program managers generally have more of an understanding of technology rather than a command of technology but must have excellent business skills related to customer interfacing, supply chain management, strategic planning, and interpersonal skills for team building.

Defining Success

There can be several definitions of success. Project managers focus on certainty by beginning projects with well-defined requirements and a clear understanding of exactly what deliverables must be provided for the customer. Success is then defined as providing these deliverables within the constraints of time, cost, and scope.

Program managers must often deal with high levels of uncertainty in the requirements, changing customer needs, possibly a continuous emergence of new business risks, and changes in technology. Therefore, it is difficult to measure program success in just time, cost, and scope. Program managers realize that their definition of success must be future oriented and view success in factors related to the long-term benefits and value that the program brings to the organization.

Re-engineering Efforts and Change Management

Organizational change is inevitable. Sooner or later, all companies undergo changes, some more often than others. The changes usually result from project successes and failures and can be small improvements to the organization's project management processes, forms, guidelines, checklists, and templates, or they can be major re-engineering efforts that impact the organization's business model. The changes can also affect just one project or all the projects within a program.

There must exist a valid justification for the changes. Some companies expend countless dollars on changes and yet fail to achieve the desired results related to their strategic imperatives. Lack of employee buy-in is often a major cause of concern.

Project managers seldom take the lead in implementing re-engineering efforts other than for the rare situation where the impact affects just one project. If the change impacts several projects that are not connected to a specific program, the leader of the change effort may be a steering committee. Program managers, on the other hand, must function as the change manager especially when the changes may impact many or all the projects within their program. Re-engineering efforts can impact just one program within a company based upon strategic program initiatives necessary.

Perhaps the most important responsibility of the person leading the change is communicating the business need for the change. Project managers may expect senior management or an executive steering committee to assume this role especially if the impact on their project is minimal. Program managers cannot and must not abdicate this responsibility to others especially if there could be a significant impact on the program's deliverables, customer and stakeholder expectations, and long-term financial considerations. Programs managers must assume the leadership role for program re-engineering efforts.

Face-to-face communication is very valuable, especially to get workers buy-in. Workers are always concerned as to how the changes will affect their job and whether new skills will be needed. People are fearful of being removed from their comfort zone especially if their vision of the future is uncertain. Project managers often have little regard for workers' concerns resulting from change management activities because the project manager's involvement with the workers may end when the project is completed, and the PM may never work with these individuals again.

Workers assigned to individual projects may have the option of requesting reassignment to other projects where they can remain within their comfort zone. Workers assigned to programs may be working on several projects within the program and reassignment may be impossible. Part of the program manager's face-to-face communications must include discussions concerning (i) the business need for the change, (ii) how and when the change will take place, and (iii) his/her expectations of the workers after the change occurs. Workers are more likely to respond favorably to the changes if the information is provided by the program manager face-to-face, or directly, rather than from someone not directly affiliated with the program.

Career Advancement Opportunities

Everybody seeks the opportunity for career advancement. In project management books and training courses, we stress that one of the roles of effective project management leadership is to help workers improve and advance their career opportunities. Unfortunately, this is easier said than done in traditional project management. Some reasons for this include:

- PMs may have little or no authority over the workers and cannot hire or fire.
- PMs may not have the authority to conduct official performance reviews and may not be asked to provide recommendations to the functional managers on how well or poorly a worker performed.
- PMs may not possess the technical knowledge needed to evaluate worker performance.
- Workers may be assigned to multiple projects under the guidance of several project managers and each of the PMs may not have sufficient time to evaluate worker performance.
- Project budgets may not have funding for the training workers need to advance their careers unless the training is specifically related to the project they are currently working on.

In most organizations, PMs do not conduct formal performance reviews. Worker performance improves when the worker exhibits his/her performance under the eyes of the person performing his/her review. If the worker on a project receives conflicting instructions from the project manager and his/her functional manager, the worker usually goes with the person conducting his/her performance review, which is usually the functional manager.

Program managers usually do not have all the restrictions mentioned above. Workers may be assigned full-time on several projects within the program, and the program manager may be authorized to make a significant contribution to the worker's performance review process. Program managers generally get to know the workers on their programs better than the project managers responsible for a single traditional or operational project.

Another important factor is the relationship between strategy and management/career development. Business strategies are formulated and executed through projects. Project management books and articles are now being written that identify the importance of aligning projects to business strategies. Why work on a project that is not aligned to one or more strategic business objectives?

There must exist a line-of-sight between senior management and project teams whereby team members understand the linkage to and importance of the strategic business objectives. Generally speaking, the linkage is often more evident between business strategy and programs than between business strategy and individual projects.

If a program has a long-term strategy and the workers recognize this strategy, then the program manager may be able to motivate the workers when they recognize the advancement opportunities through the business strategy for the projects within the program. Programs with growth strategies generally offer workers more management development opportunities than functional organizations that cater to single projects and focus on stability and possibly retrenchment strategies. Also, the re-engineering efforts discussed previous, if explained carefully to the workers, may identify career advancement opportunities.

Data-Driven Risk Management

An important lesson learned from the COVID-19 pandemic was that traditional approaches to risk management may be ineffective during a crisis. The risk associated with a failed project may be inconsequential compared to the failure of a long-term program.

Program managers require significantly more data than project managers for risk assessment since many of the projects within the program may be strategic rather than traditional or operational projects. Data-driven risk management will require access to information warehouses and business intelligence systems. As stated by the authors (Kerzner and Zeitoun 2022):

> *The changes that have been taking place in business and in the way of working of programs/projects have led to an unprecedented level of uncertainty that make the topic of estimating and the associated risks central to the success of the strategic initiatives. It is in our view that digitally enabled estimating requires innovation in order to create a commercially successful product, which also means that the team members must understand the knowledge needed in the commercialization life cycle starting from the early projects' stages.*

Some project managers might simply walk away from a failing project and then move on to their next assignment. Program managers have significantly more at stake and focus on ways to salvage as much business value as possible.

Risk management is often looked at differently whether seen through the eyes of a project or program manager. Project managers tend to focus on negative risks, namely the likelihood of something bad happening and the resulting consequences. The intent is to reduce negative risk and the ways we have taught it according to the PMBOK Guide are with strategies to avoid it, transfer it, mitigate it, or accept it.

Program managers must deal with positive risks as well as negative risks. Positive risks are opportunities to increase the business benefits and business value of the activities within the program based upon what is in the best interest of both the parent company as well as its customers. Opportunistic strategies include accepting it, exploiting it, transferring it, and enhancing it. Effective risk management activities for programs must be holistic in nature and consider both negative and positive risk management strategies.

Stakeholder Relations Management

Stakeholder relations management affects key principles in the PMBOK Guide. For the project manager, stakeholder relations management focuses heavily upon providing project performance feedback to stakeholders and engaging them in decisions and execution of the project plan. For program managers, the relationship is more complex.

Most project managers do not have the responsibility for marketing and selling of the deliverables of their projects. Project managers move on to their next assignment after project closure regardless of the business value of the deliverables and might never work with the same customers or stakeholders again. Program managers view themselves as managing a portfolio of projects with the strategic intent of creating a long and profitable life expectancy for the deliverables created within the program. Therefore, program managers may need much more sales and communication skills than project managers. This requires more frequent communications with customers and stakeholders.

Both project and program managers have a vested interest in quality management, but often for different reasons. For the project manager, quality is most often aligned with the organization's Customer Relations Management (CRM) program which looks for ways to sell more of the existing deliverables to the customers in the short-term. The focus is on short-term thinking and quick profits.

Program managers have more of a long-term and strategic perspective by focusing on Customer Value Management (CVM) rather than CRM efforts. The intent of CVM is to get close to your customers to understand their perception and definition of value and what value characteristics will be important to them in the future. This allows the program managers to create meaningful strategic objectives aligned to their customer base. It provides a much closer and stronger relationship with important customers. This can lead to lifetime bonding with critical customers and create a sustainable competitive advantage.

Multiproject Management and Innovation

Traditional project management focuses on managing a single project within the constraints of time, cost, scope, and quality. Innovation requirements are usually not included in most traditional or operational projects. The emphasis is usually on the application of existing knowledge and technology in the creation of deliverables. Program managers are required to perform in a multiproject environment using many of the concepts required for effective portfolio management practices. Multiproject management and effective customer relations management practices give program managers more innovation opportunities than project managers working on a single traditional or operational project.

Some of the issues that program managers must address come from answers to the following questions:

- How well are the projects within the program aligned to strategic objectives?
- Must any of the projects be cancelled, consolidated, or replaced?
- Must any of the projects be accelerated or decelerated?
- Does the portfolio of projects need to be rebalanced?
- Can we verify that organizational value is being created?
- How well are the risks being mitigated?

Answering these questions requires the program manager to utilize business and strategic metrics for informed decision-making regarding strategic opportunities, new technologies that may be needed, and new products/services customers expect. Strategic opportunities have a strong linkage to innovation activities that address the customers' definition of value.

Previously we stated that program managers have a greater interest in CVM than CRM and therefore must address innovation opportunities in the multiproject environment. Program managers must recognize that innovation-based strategies can become the key drivers to maintain or create a sustainable competitive advantage. By managing all the projects within the program in a coordinated manner whereby each project may be related to other projects, program managers are able to create significantly more business benefits and business value than traditional projects managers. Obtaining this synergy requires significantly more metrics and some expertise in innovation practices necessary to support CVM activities.

WHY THIS BOOK

As I was doing my undergraduate degree in civil engineering, a dear professor of mine introduced us to the mechanics of construction management. It was a fascinating experience that opened my eyes to a new and a different world of management at the time. Between breaking down the project into an organized set of elements and learning the steps of figuring out the critical path that determines a given project duration, I was mesmerized and thought this is such a great field of practice. Although I started my career as a practicing engineer, designing city and county utilities for example, I always had fond memories and great learning from this encounter with Prof. Hosny.

It was natural to get my Project Management Professional (PMP), and I was very proud to be the first certified professional in the city of Wichita, Kansas, USA, where I lived at the time. The biggest aha moment that I experienced then, was when I shifted from engineering to manufacturing to run a PMO for the first time. The realization that the project management practices apply almost equally well across most industries was a beautiful discovery. One thing I found then and continued throughout all my career changes and milestones was "projects could be thought of as successful, yet the impact envisioned or initially perceived by the customer or user of the service/product of the project might not have been realized."

This was the reason why I started to investigate and understand program management principles and got ever more excited about that field and its importance to creating the most meaningful changes. Programs are strategy execution vehicles and this means that the intense focus on benefits management comes as a key ingredient to the inception of a program and the managing of its lifecycle. This would get us over this challenge of winning the battle and losing the war that could happen in project work that might be focused on just getting deliverables done, all the while missing the achievement of critical stakeholders' outcomes.

APPROACH

Throughout this textbook, the focus will span both the program management processes and techniques as the vehicle to achieving change and the person whom we give that title and responsibility of being the program manager. You will learn how to design your programs for success, how to prioritize and focus on achieving value, and most importantly develop the right skillset and qualities needed to lead the future most important change creation efforts. The principles presented and analyzed throughout the text align with many of the global industry standards, with closer alignment to the Project Management Institute (PMI) standards and practice guides.

The orientation of the textbook is in the direction of the practitioner and thus you will see cases and examples targeting how to apply program management principles and how to overcome the gaps and resistance that might be faced when working across organizational boundaries and the multiple program stakeholders and their competing demands. Yet, a few of the readers might also be candidates for pursuing professional certifications in the field of project and program management. There will be useful references throughout to guide these readers to the resources most relevant to these certifications too. In addition, there is an instructor's manual for the book and faculty can obtain this manual by contacting John Wiley Publishers.

THE EXPECTED OUTCOMES

It is my hope as the author that the reader of this textbook will develop a strong appreciation of the value of program management in enabling change. Understanding and practicing how growing beyond project management to the land of strategic value of projects in the form of programs is a critical outcome of this work. As a future leader of programs, the qualities that you develop, and the related critical skills necessary to achieve strategic benefits, should empower you to consistently tackle the most challenging transformation initiatives.

In an increasingly digitally driven universe, it is the strong mix of human behaviors and stakeholders' management practices addressed in this textbook that is the true difference makers for what success would look like in future organizations. The outcome of this textbook should be customized to the needs of the individual practitioner, the maturity of his/her project/program team, and the stage of the organization in its understanding journey of the value of projects and programs.

As the leader of future programs, learning to sharpen your adaptability in articulating and driving the story of your program and projects will dictate your success and your ability to make dreams a reality. Leading programs in the world from the angle of the prevailing VUCA (Volatility, Uncertainty, Complexity, and Ambiguity) is challenging enough. Yet as we mature this to the likely prevailing environmental versions for this coming decade, or the many others that might encounter our future path, managing programs will become even more challenging with an equally high potential for opportunity.

This could be a sandwiched VUCA between the two Ds of Diversity and Disruption (D-VUCAD) or the more likely future business state of Dynamic, Ambiguous, Uncertain, Nonlinear, Complex, and Emergent (DANCE). The key is that the program manager will have to more frequently and effectively call the shots, operate as an entrepreneur, make tough decisions, continue to connect vast variety of diverse stakeholders, drive integration, take Artificial Intelligence's impact into the mix, and ensure that every day in the life of a program is focused on achieving value. I would like to think of this as the ***Program Way*** in a future that is dominated by sensing, responding, and capitalizing on strategic business opportunities.

Wishing you a reflective read and the very best on your own journey of excelling in delivering consistent and sustainable value from your projects and programs.

REFERENCES

Kerzner, H. and Zeitoun, A. (2022). The digitally enabled estimating enhancements: the great project management accelerator series. *PM World Journal* XI (V).

Project Management Institute (2021). *A Guide to the Project Management Body of Knowledge (PMBOK® guide)*, 7e.

Weaver, P. (2010). *Understanding Programs and Projects—Oh, There's a Difference! Paper presented at PMI® Global Congress 2010—Asia Pacific, Melbourne, Victoria, Australia*. Newtown Square, PA: Project Management Institute.

SECTION I

GOVERNING WITH EXCELLENCE AND ACHIEVING CHANGE

SECTION OVERVIEW

This section describes the fundamentals of program management. The value of program management in achieving organization change is detailed. A few initial tools to aid the program manager in achieving clarity and leadership of the change journey are explained. Governing with excellence is supported with a few case studies across global organizations.

SECTION LEARNINGS

- The relationships among portfolio, programs, and projects to understand how they align to deliver value.
- Why programs and strategy have to be connected to achieve meaningful success and gain a competitive advantage.
- The linkage process between program management and change management and its use to drive and sustain innovative outcomes.
- How the different program roles support its success, in addition to leveraging the role of the program sponsor in closely connecting to, and achievement of, business goals.
- Using a program roadmap to create the connecting blueprint that integrates the focus and expertise of the program team.

KEY WORDS

- Strategy
- Change management
- Complexity
- Program sponsor
- Excellence
- Roadmap

Program Management: Going Beyond Project Management to Enable Value-Driven Change, First Edition. Al Zeitoun.

INTRODUCTION

The world must continually change to sustain humanity and its needs and to address the strategic complexities we all encounter. The execution of strategies in this changing world of work has been going through an evolution from its normal way of working. This transition is only going to be faster and more complex. As a leader and a program manager, it is becoming highly critical to learn and develop governing strategic and integration capabilities and possess a crystal-clear focus on value. This section highlights the challenging and rewarding journey towards program delivery excellence. It stretches the learners' skills of creating, thinking, and reimagining futures that are becoming more likely scenarios. You will learn how to deliver faster and remove strategy execution friction.

This work will build on the principles of the Project Management Institute's (PMI) program management standards. Of importance will be the focus on bringing the power of integration as the engine for programs success. This will help you drive and support meaningful change and realize a strong return on your programs' investments by deep diving into a selection of practical capabilities.

Building the necessary integration capabilities requires time and energy in exploring what it takes to stretch your leadership style, while uncovering the processes and principles of program management as the vehicle for creating change. This learning journey is designed with a global perspective in mind and is intended to help accelerate a project practitioner's growth as a future leader and a successful program manager.

Across the chapters of this section, there is an opportunity to learn and understand the delicate balance between program management and change management. A multitude of program management mechanisms provide the depth opportunity needed to drive achieving change. Programs by nature tend to also have a long-time frame, where multiple changes could occur. Leaders need to see programs as strategic business opportunities that are designed to create the organizational strategic objectives and/or solve critical customer needs.

There will be several global case studies to be addressed that propel us into the future of data and its role in program delivery, innovative decision-making tools, and the appreciation of the human side of change. An excellence model will also be referenced as a good anchor for the many behaviors that are critical to learn and practice across the planning and execution of programs.

A program manager needs to fully immerse in a continual learning lifestyle. As the leader of a program, she/he has access to a wide variety of learning instruments and success will depend on how the leader takes the time to step back and think about the learned practices and then personalize that to your specific learning and program work needs.

There are potentially different future roles in the growing program management field. Examples highlighted in this section will cover the conductor, the change maker, the holistic leader, and the storyteller. This provides the program leaders an opportunity to experiment with the most fitting role and where a clear impact is best created. These, and many possible others, are all roles you could practice, practice, practice. What could be more exciting than being equipped to lead, or participate in, the world's most meaningful programs?

This section is built on a global perspective to bring life to the principles and practices of program management. This should enable you and your organizations to experience the potential this program management discipline brings to the way you and your teams work now and into the future.

Before we get into the first chapter, let's highlight some of the typical challenges faced by programs and that affect the successful management of these critical change efforts.

PROGRAM MANAGEMENT CHALLENGES

A program is generally defined as a grouping of projects that can be managed consecutively or concurrently or a combination of both. There are numerous challenges facing the program manager that quite often makes it difficult to achieve all or even part of the strategic goals and objectives established by senior management. The larger the program, the more difficult it will be to overcome the challenges.

Many of the challenges are common to both projects and programs. However, the risks due to the challenges may have a much greater impact on programs than projects. When projects are challenged, some companies simply let the project fail and the team moves on to their next project assignment. When programs are challenged, the cost of terminating a program can be quite large and have a serious impact on the organization's competitiveness and future success.

Projects generally have a finite time duration. Most programs, because of their strategic nature and impact on the success of the organization, are much longer in duration and are susceptible to more challenges, risks, and negative impact on the business.

TYPES OF PROJECTS

When managing individual projects that are not associated with a program, we normally treat them as "traditional" or "operational" projects. Program management activities are composed of "traditional" as well as "strategic" projects. Traditional projects begin with well-defined requirements, a detailed business case, a statement of work, and possibly a Work Breakdown Structure (WBS) that outlines all activities needed for the duration of the project. Strategic projects can begin based upon just an idea and the requirements are elaborated upon as the program progresses.

The mix of traditional and strategic projects varies with each program. The business environment seems to be changing and more programs are becoming strategically oriented. Reasons for the strategic orientation include:

- Today's business environment has much greater risks and uncertainties.
- There is a growing need for more products, services, markets, and customers.
- There is a much greater need for internal projects that promote better efficiency and effectiveness.

The challenge will be to accept more strategic projects as part of a program and recognize that most of these strategic projects cannot be managed using the traditional project management processes, tools, and techniques that have been taught for decades.

PROGRAM BUSINESS CASE

Many of the projects within the program will be strategically oriented and will be initiated without detailed requirements or even a business case. However, the overall program must be based upon a well-understood business case because of the cost and time associated with a program.

The business case must articulate the expected benefits and business value expected. The business case also provides the boundaries for many of the decisions that will have to be made. The challenge will be in the preparation of the business case such that all program team members clearly understand what is expected of them.

SCOPE CREEP

The risks associated with scope creep can occur on almost all programs and projects. However, the magnitude of the risks is greater on programs. The magnitude of the risk on programs depends upon the relationships between projects within the program. Scope creep on one project could cause costly scope changes on other projects within the program as well.

Change Control Boards (CCBs) are often created to deal the with approval or denial of a scope change. For projects, the CCB may be composed of project team members, organizational management, and stakeholders impacted by the change. On programs, determining the numbers of stakeholders impacted may be difficult. Decisions made by CCBs on programs must consider the impact that the scope change will have on the entire program rather than just on one project.

Another important consideration on programs is the impact that change might have on suppliers and distributors. The cost associated with bringing on board new suppliers or distributors could create a financial headache on programs.

ORGANIZATIONAL CHARTS

People can be assigned to projects or programs full time or part time based upon the size and/or duration of the activity. Projects tend to be shorter in duration than programs and usually do not have an organizational chart unless required by the contract. Projects use Responsibility Assignment Matrices (RAMs) and the only people that might appear on an organizational chart would be the project manager and assistant project managers, if applicable. Other workers, such as from sales, engineering, procurement, legal, and manufacturing, appear as dotted lines to the project manager and solid lines to the functional area.

Programs can have detailed organizational charts with full-time assignments from people in sales, engineering, and manufacturing. Because of the number of projects within a program, full-time sales personnel may be assigned for projects involving competitive bidding where proposals must be prepared continuously. The same may hold true for engineering and manufacturing personnel that will appear full time on the organizational chart and reporting directly to the program manager. But in reality, they may still be dotted line reporting to the program manager and solid line reporting to their functional managers regardless how it appears on the program organizational chart. The intent of showing reporting relationships, either solid or dotted, is to help identify the right accountability and eliminate the risks of mismatched skills.

Large programs may have special requirements that justify the need for full-time personnel. Examples might include:

- **Program office manager:** This can include handling administrative paperwork, meeting scheduling, and making sure that program activities are aligned to company standards and expectations.
- **Reports manager:** This person is responsible for the preparation and distribution of all reports and handouts. The person is usually not involved in the actual writing of the reports.
- **Risk manager:** This person monitors the VUCA environment, and the enterprise environment factors. Additional responsibilities include risk identification, analysis, and response to all risks that can impact program success.
- **Business analyst:** This person works closely with the risk manager and activities may include identification of business opportunities and threats. The analyst may monitor compliance to customer requirements and verification of the program's deliverables.
- **Change manager:** Some large programs may clearly indicate that changes in the firm's business model will be necessary. The change manager prepares the organization for the expected change. The change may just be in some of the processes or the way that the firm conducts its business rather than a significant change to the business model.

Removing critical resources from their functional area and assigning them full time to the program manager may seem like a good idea. However, there are risks if the program comes to an end and the workers do not have a "home" to return to.

Some people look at a program's organizational chart and believe that there are too many people, and the program is overmanaged. But we must remember that there are a multitude of points of contact on programs, all requiring possible continuous interfacing. The alternative would be poor communications where certain stakeholders might be left in the dark.

MANAGING STAKEHOLDER EXPECTATIONS

Stakeholders are the people that ultimately decide whether a program is successful. There can be significantly more stakeholders on programs than projects. Failing to meet program stakeholder expectations can result in a significant loss of business. Given the long-time frame of many programs, managing the changes in stakeholders over the lifecycle of the program and addressing their changing expectations are important muscles to develop in this ***Program Way***.

On projects, especially if short duration, project managers may be able to keep stakeholders at arm's length and minimize their involvement in the project. Program managers may not have this ability because of the power that some stakeholders possess. Managing customer expectations is a continuous process, beginning at the start of the project, and can help deal with stakeholders that continuously change scope because they cannot make up their mind as to what is really needed.

Program managers must ensure that program performance is aligned to the key strategic objectives of the stakeholders. This requires continuous rather than intermittent communications of performance and the ability to quickly address scope change requests and the resulting consequences.

Managing stakeholder expectations can become difficult if the stakeholders span several time zones. This can limit face-to-face interactions between stakeholders and the program team. The situation can be further complicated if many of the stakeholders have different cultures that can come with different areas of interest and levels of importance when resolving program issues.

STATUS REPORTING

Stakeholders want to see facts and evidence that program progress is being made toward benefits and deliverables expected. Monitoring and controls systems that are used on traditional projects may be insufficient for reporting program status. The program manager may need additional metrics that are not included in the Earned Value Measurement System (EVMS). Additional information required by stakeholders may necessitate the creation of business metrics, value and benefits tracking metrics, intangible metrics that measure the effectiveness of activities such as executive and stakeholder support, and strategic metrics that measure the alignment of projects to strategic business objectives.

Another issue is that many program managers follow traditional status reporting guidelines on the frequency and timing of the reports whereas stakeholders may want the information more frequently. Some programs use dashboard reporting systems where the dashboards may be updated as often as on a daily basis.

The format of the status report can also be an issue. Project managers may be content with providing status using a standard template. For programs that use dashboard reporting systems, program managers may find it necessary to prepare a customized dashboard for each stakeholder.

STRATEGIC ALIGNMENT

An important function of status reporting is to identify strategic alignment. Identifying strategic alignment requires that a line-of-sight be established between project teams and senior management. The concept of line-of-sight is now taking place in most companies as part of standard project management practices. Line-of-sight provides guidance to project teams to ensure that decisions are being made such that projects remain aligned to organizational strategic business objectives.

On programs, strategic alignment is more difficult because the program manager must align the projects within the program to his organization's strategic business objectives as well as the strategic objectives provided by the stakeholders. Conflicting strategic objectives will generate a lack of proper high-level alignment resulting in conflicts among project teams as to what are the real constraints and boundary conditions of the projects within the program.

SELECTING A METHODOLOGY

For decades, companies have endeavored to create a one-size-fits-all methodology that could satisfy the needs of all their projects. This approach worked well as long as the focus was on traditional or operational projects that had some degree of commonality.

As we began applying project management practices to other types of projects, such as those involving strategic issues, innovation, and R&D, the need for flexible methodologies such as Agile and Scrum became apparent. This created good news for standard project management practices, but challenges for program management.

Project managers selected methodologies that were aligned to their customer's business model. In many cases, this included using the customer's life cycle phases and gates. The selection of a methodology that could be aligned to the customer's way of conducting business brought with it high levels of customer satisfaction and repeat business.

Program managers could not follow the same path because of the number of customers and the complexities that this would create for progress reporting. Program managers will be allowed to use multiple methodologies for customer strategic alignment purposes, but not necessarily a unique approach for each customer.

DEFINING SUCCESS

Project and program managers can have different definitions of success. Most project managers working on traditional projects define success in terms of deliverable created and short-term business benefits and value. Program managers are long-term thinkers and may measure success in terms of long-term benefits and value achieved as well as repeat business.

The current trend on projects is that the project manager works closely with the customer at project initiation to come to an agreeable definition of project success. The project manager, possibly accompanied by the customer, will then select the metrics to be used on the project to measure and report progress toward the agreed definition of success. This can include metrics specific to an individual customer.

Programs become very difficult to manage when there are multiple definitions of success, especially if each success definition requires different metrics. Program success is measured in business terms usually, rather than technical terms, with a focus on a sustainable competitive advantage through business value creation.

CHAPTER 1

Connecting to Purpose and Achieving Change

This first chapter is focused on the understanding of the typical program's ecosystem. It is intended to highlight foundational program management principles that help the program manager to clarity the direction, align the key stakeholders, and design the program plans with the customer in mind. This is intended to enable setting the stage for the ***Program Way*** of envisioning the program and its components and set the execution steps on the right track.

In a fast changing world, where Artificial Intelligence applications usage increases, integrating connecting prevails, and cloud applications become the norm, the program managers become the future architects of change.

1.1 PROGRAMS MATTER

Key Learnings

- Understand the value of programs in the delivering strategic outcomes
- Get introduced to an excellence model to help you become more aligned in your program work
- Explore a case and an estimating tool focused on driving program decisions for speed and quality
- Learn how to develop the holistic views that prepare you to exercise the role of the program manager as the conductor
- Address complexity for better, customer-centered solutions

Project and program management have been maturing rapidly over the last decade. The signs of that are seen everywhere, whether in the increasing number of global events in the field, the awards dedicated to recognizing the impact of program and project managers, the refreshed value or program offices, and ultimately the improving business leaders understanding of that value programs strategically bring to their organizations.

The nature of programs is unique. The idea that one has an overarching connecting structure to its individual components, being sub-programs or projects, is a valuable one in creating the often-lost holistic visibility. Programs teach us to be strategic, they enable us to build capabilities and develop muscles that would otherwise be neglected without the discipline programs require and bring.

Program Management: Going Beyond Project Management to Enable Value-Driven Change, First Edition. Al Zeitoun.

The simplest way to highlight the difference between a Program and a Project:

- Project = beginning and end with a focus on deliverables
- Program = connects the sub-programs and projects, continues this on-going connectedness, still could have a targeted end in sight, and is highly linked to strategic purpose the organization has for this integrated strategy execution vehicle

What makes the program manager role most unique is having to develop a suite of skills that differ or build on what a project manager must possess and that center on the need to constantly wear the hat of value. This is the link to the business cases behind the program and ultimately what one would refer to as the "so what?" that could be missed in the noise of high deliverables focus that prevails in the classical project environment. This heightened Business Value matters strategically as it allows the program manager and the program team to drive the entire cross-organizational engine toward value from the get go, rather than catching that later in the program journey at a much higher price.

One of the critical reasons why programs matter is the fact that stakeholders' management and engagement is a key enabler for programs' ultimate success. Program managers invest in being politically savvy and in creating the comprehensive understanding of the stakeholders that allows them to develop the right engaging and influencing strategies. The groupings of stakeholder into buckets, such as interested, involved, and impacted, ensure that the right bets are placed on who is going to contribute what and how to a program's lifecycle and thus directly links to achieving a given targeted program's business value.

1.1.1 The Siemens Case

Understanding programs' complexity and finding ways to simplify it is critical. The changes that have been witnessed over the past decade regarding the imparlance and value of data have been immense. It is to be expected that digital innovations, and the associated data analytics and intelligence, will continue to dominate the next decade. Using case studies across global organizations help us shed the light on how to best break organizational barriers and create the right transformations in how we work and become more effective in achieving programs' successful outcomes.

In this Siemens case, highlighted in the whitepaper,[1] Zeitoun 2021, a combination of the strength of simulation and the clarity of an integrated view of a program lifecycle gives us the edge in enhanced prioritization at the core and showcases how to achieve the critical focus shift to outcomes as needed in programs (especially what matters the most to customers).

The role of data in creating programs value is multiplying. Simultaneous access to data via groups of stakeholders is critical to programs' success. This allows the envisioning of the program's roadmap, and the associated critical decisions, to become easier and more effective. Achieving quality decisions is enhanced with solid prioritization capabilities that include a dynamic re-prioritization capability that is becoming a must-have in the continually agile delivery of programs across organizations. Data seamless access allows us to achieve higher adaptability to changing program conditions and related external environment fluctuations.

As highlighted in Figure 1.1, the aforementioned elements are all pieces of the puzzle of the future digitalized program management. Program management is changing due to the degree of digital enablement. This is leading to a complete change of the way of working, collaborating, and most critically where the program mangers spend their time to create the most impact.

The figure also shows the importance of sharpening the focus on outcomes and that leadership component, which has become a key to the role of program manager, more than just outputs' focus. Additionally, streamlining customer's inputs is becoming a top agenda of most organizations in their pursuit of the customer centricity expected in today's and future businesses. This is achieved with the close mapping to the customers, their changing needs, their direct voices in the development and production process and establishing ongoing clarity through the noise.

---

[1]Siemens Whitepaper – https://resources.sw.siemens.com/es-ES/white-paper-integrated-program-lifecycle-management-consumer-product.

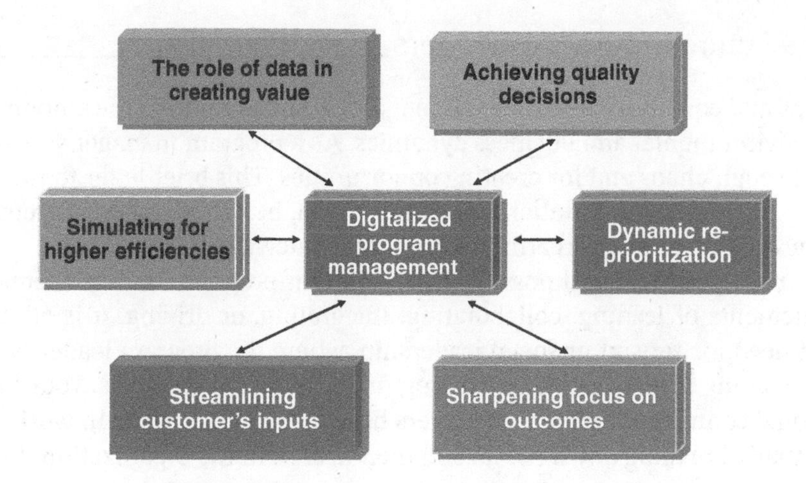

FIGURE 1.1 Digitalized Program Management Effects

Speed is an expected way of the future and simulating the work of a program, and envisioning the entire program's roadmap is a great way to predict and prepare for risks and thus contribute to higher Efficiencies. Digitalized program management enhances efficiencies in addition to the previous effectiveness points highlighted in this case.

1.1.2 Connecting to Strategy

Strategy is difficult. Being strategic is an organizational muscle that requires continued investment in its development. Strategy is an articulation of the aspirations of the organization and the intended building blocks of the investments and work required to get there. Strategic Aspirations of an organization or a team are typically broken down into buckets of work (programs and projects) to be achieved in the form of certain outcomes over time horizons.

Figure 1.2 shows the bi-directional impact program benefits have on strategic aspirations in addition to the envisioning of programs and related initiatives. The figure also reflects the role of strategic objective in maintaining the strategic clarity and continually updating what strategic success looks like, while reflecting that in necessary directional changes.

Programs Benefits are the link to strategy. For that reason, the lifecycle of a program needs to be benefits-focused. A natural cascade takes place to allow for capturing the benefits and their correlation to achieving strategic outcomes. Programs and related Initiatives form a map of connectedness, via seeing all programs in an organizational portfolio and their exact contributions to strategy. Another valuable aspect of digitalized program management is strengthening this visualization of the interplay across programs. Strategic objectives are the other strategic ingredient in the program manager's arsenal. They provide the holistic views that drive program success and support the integrator role of the program manager.

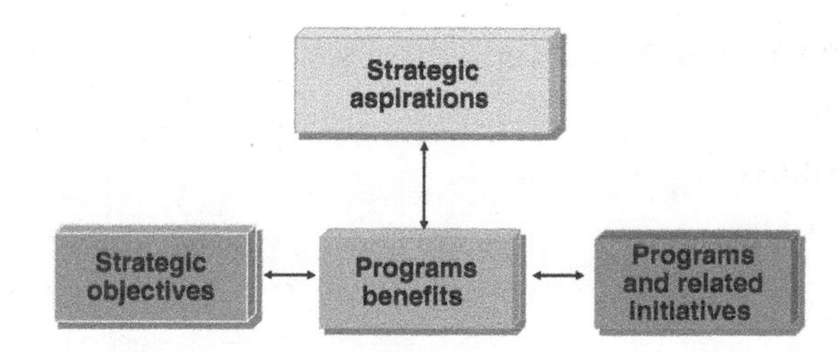

FIGURE 1.2 Program Benefits Strategic Links

1.1.3 Letter to the Future Program Managers

There are both concerns and equally a level of excitement about the future with the amount of disruption and the anticipated changing environmental and business dynamics. As a program manager you are in the right place at the center of leading through chaos and for creating opportunities. This brief letter to future program managers is intended to highlight some of the key anticipated shifts ahead, be aware of them, prepare for them, and ultimately put that readiness to good use in creating meaningful strategic impact.

The changing nature of your role and possibly title. Whether program manager remains as a title, or it get replaced with some elements of leading, collaborating, integrating, or driving, it is all about creating impact. There is a dominating need for servant or social leadership, where the program leader is able to adapt between being the coach and becoming the one carrying the program team across obstacles. Your future role is shaping to be the true organizational connector. Program managers have to have their voice in working across the business boundaries and continually breaking down actual and mental silos in the organization. Future program managers are connectors.

Readiness for that future also has growth and people components. The open window for continual learning is a feature that is strengthened by technology and artificial intelligence. This requires developing an appetite for your growth and for equally growing other key stakeholders around you. A more mature and developed stakeholder community directly contributes to making your future role most effective.

In the future, program managers will also have a vital impact on sustaining the growth of business and people. With the norm shifting to program managers being more aligned with the executive teams, being part of the most critical strategic dialogues and decisions, and having the right seat at the table, the value of program management continues to become more evident. This makes your role even more clearly strategic in terms of impact driving and affecting the future of organizations.

1.1.4 Strategy Execution and People

A well-planned strategy demands a clear roadmap for executing that strategy. This visualization of a strategic journey requires the collaboration of a large group of stakeholders, the right set of data, and an organized way to make sense of all the details. Program management's success is about connecting to the source, the strategy. Executing a program without a clear line of sight is a recipe for challenges, misalignment, and for potentially missing the mark on the right meaningful outcomes.

Strategy in the upcoming decade is expected to continue becoming more fluid. It is still going to be designed as a combination of ideas, aspirations, build on key discussions of stakeholders, and need programs as vehicles for getting from A to Z. Programs are critical for intentionally driving changes along the way and they use roadmaps to show the path forward.

People and the way they work are at the core of effective execution of these strategies. This is built on clarity of the strategy's purpose, effective communication of that purpose, and clarity of who is doing what.

Guided collaboration supports achieving people role clarity and directly contributes to the success of the Program Manager. This results in a number of important outcomes:

- Enables connecting the dots
- Integrates across stakeholders and their diverse interests
- Aligns across the business
- Achieves value to the business

Review Questions

Parentheses ( ) are used for Multiple Choice, when one answer is correct. Brackets [ ] are used for Multiple Answer, when many answers are correct.

Why does the role of a program manager matter to achievement of business outcomes?

( ) Creates a handful of metrics to achieve control.
( ) Builds and integrates cross programs' focus on realizing benefits.
( ) Engages extensively with program stakeholders.
( ) Ensures executive team looks good.

How does digital directly help with enhancing programs' customer-centricity? Choose all that apply.

[ ] Sharpens the focus on outcomes.
[ ] Expands the role of data in creating value.
[ ] Supports knowledge-management infrastructure.
[ ] Streamlines the customer's inputs.

Where do future program managers need to focus most to remain relevant?

( ) Getting done on time and within budget.
( ) Expanding tighter governance.
( ) Ensuring alignment of program with strategic drivers.
( ) Strengthening conflict resolution skills.

1.2 ALIGNMENT ACROSS DELIVERY

Although most companies understand the importance of achieving benefits from their investments, there is a possible worst scenario related to getting work done, resources used, and budget spent and then not realizing benefits. This late discovery could result in a multitude of unpleasant results, from the classic blaming exercise that could take place, to real consequences for the program team, or even external results that affect the viability of the organization with its customers and other key stakeholders. If positively handled, it would be a learning opportunity where the organization could rethink its program choice, its processes, and the supporting resourcing of people and technology.

Key Learnings

- Understand the obstacles to benefits realization
- Linkages across: Focus on alignment across the 3Ps
  - Portfolios
  - Programs
  - Projects
- Explore program lifecycle phases to create a comprehensive understanding of the program
- Link to excellence by exploring the EFQM model (dynamics related to program management)
- Common reasons for tailoring delivery approach (why is this important and why one approach across program(s) does not work)

1.2.1 The Worst Scenario – Not Realizing Benefits

1.2.1.1 The Six Key Obstacles to Benefits Realization

Obstacle #1 Stakeholders Involvement

There is no targeted and active involvement of key stakeholders.

- Key to the success of your role as a program manager
- You won't get the right focus without the proper engagement

> **TIP**
>
> Invest in building the relation with key stakeholders and find ways to pulse what works best for their involvement in supporting your program's direction.

Obstacle #2 Missing Benefits Plan

The project or program is approved without a well-developed benefits realization plan.

- Not investing in developing one
- Who owns it?

> **TIP**
>
> Focus on knowing what the projects within the program contribute to achieving its benefits.

Obstacle #3 Deliverables Emphasis

There is a heavy focus on the projects' deliverables rather than on the creation of business value.

- Could be a negative thing
- Outcomes–orientation is what counts most

Another opportunity for aligning outputs to business value

> **TIP**
>
> Ensure being continuously strategic about the reasoning behind the program and the impact it is expected to create

Obstacle #4 Success not Clearly Defined

No Alignment on Program/Project Success Definition

- Programs could fail before they start
- Not just because of not spending enough in planning
- More an issue of taking the time to align around the definition of success

> **TIP**
>
> A key contributor to success is to use the breaks early on and ask questions to ensure we have a joint view of what success looks like before we run.

Obstacle #5 No Benefits Tracking

Failing to track benefits over the end-to-end lifecycle of the program/project.

- Not having the right measures or the right dashboard
- Needs some indication that we are on or off track
- Needs to be end-to-end

> **TIP**
>
> You need an approach to benefits tracking that integrates the program's components and their contribution to benefits.

Obstacle #6 Poor Change Management

Lack of an acceptable transformational and organizational change management processes.

- Program management is a critical strategic vehicle for creating a change
- Lacking proper change management processes creates gaps in understanding the transformation outcomes expected
- Goes hand in hand with the program management processes

> **TIP**
>
> Remember programs are there to create change. Change has to be managed and supported with the relevant processes and change leadership

1.2.2 The 3Ps: Definitions

The 3 Ps are project, program, and portfolio. As per the PMBOK® Guide, Seventh Edition, a project is a temporary endeavor undertaken to create unique product/service/result. When we go to the program level, we deal with related projects. This could be in the form of subsidiary programs or components. The program also has a set of activities. These activities are to be managed in a coordinated manner. The ultimate goal here is to obtain benefits not available from managing them individually.

The portfolio on the other hand is the bucket that covers a multitude of components that summarize the organization's investments in its strategic growth. It includes projects, programs, subsidiary portfolios, and operational work

- manage as a group to achieve strategic objectives
- manage in an integrated and connected fashion
- it is the umbrella
- collectively drive what the organization does

It is critical to have clarity of how the 3Ps connect in purpose, scope, execution approach, and the ultimate achievement of the change effort.

1.2.3 Linkages Across Portfolios, Programs, and Projects

Mapping the linkages across the 3Ps is a strategic step that requires the right stakeholders to have the clarity of purpose behind the company's investments and the work necessary for the portfolio and its elements. Figure 1.3 highlights a sample properties portfolio. That portfolio could have portfolios within it. In this case a maintenance portfolio is shown and further broken down into an innovations program and also covers the related operations work.

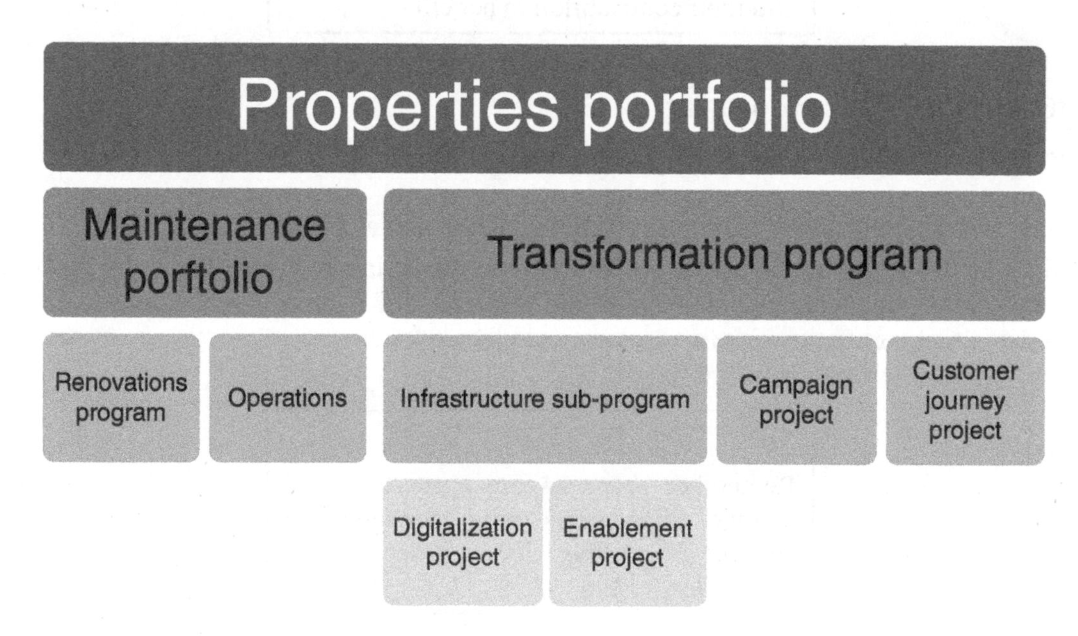

FIGURE 1.3 Sample Linkages Structure

In addition, the properties portfolio includes a transformation program that further cascades the elements of related sub-programs and projects.

> **TIP**
>
> Remember programs are there to create change. Change has to be managed and supported with the relevant processes and change leadership.

- Top level is the portfolio
- Then programs
- Sub-programs
- A cascade into levels of details
- Having that connected holistic view
- Could have a portfolio leader or manager looking at the connectedness across the portfolio elements

1.2.4 Program Lifecycle

The program lifecycle is the journey from inception to completion. The envisioning and execution of the program requires an integrated view of the path to benefits. It assumes a strategic capability in the organization that allows leadership to coalesce around the interactions necessary to make a program a true success (Figure 1.4).

- Different from a classic project lifecycle that mainly has phases and gates
- Program charter drives the journey
- The front piece is very important
- Delivery has the heavy-duty work, yet the emphasis needs to be on the proactivity and planning aspects
- Closure also depends highly on what happened on the front and what success looks like
- Programs contribute to a culture of excellence in the organization

1.2.5 Excellence and the EFQM Model

Excellence has been driving agendas of many global organizations. Alignment from the strategy level through the execution level to the results is a critical excellence enabler. A multitude of project management, change management, and quality management institutes and professional organizations have been supporting the practices that are key fundamentals to achieving that excellence.

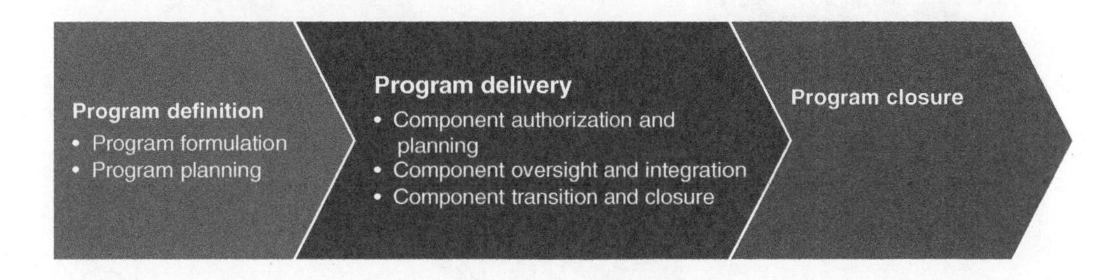

FIGURE 1.4 Program Lifecycle

The European Foundation for Quality Management (EFQM) is such an organization example. Its practices relate to programs work and delivery excellence in three distinct ways:

- aligns strategy/benefits
- power of execution
- addresses cultural/leadership/ transformational capabilities

The principles of EFQM reflect the dynamic nature of continuous improvement needed for excellence (repeatability of successful patterns, time and again, thus using the programs vehicle as a mechanism to enhance excellence). Figure 1.5 shows the high-level connection between the program's clear direction, the related necessary execution efforts, and the importance of continually calibrating that progress against meaningful results that matter to the purpose initially set in the direction part of a program's journey.

1.2.6 Tailoring of Program Management Approach

- Figure 1.6 highlights that we need inputs throughout this tailoring journey, adjusting the program reasons, returning to the focus on the customer value, and that clear context matters

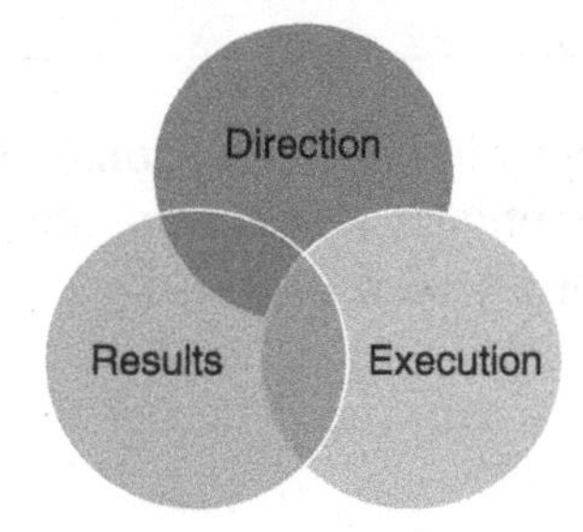

FIGURE 1.5 Program Delivery Excellence

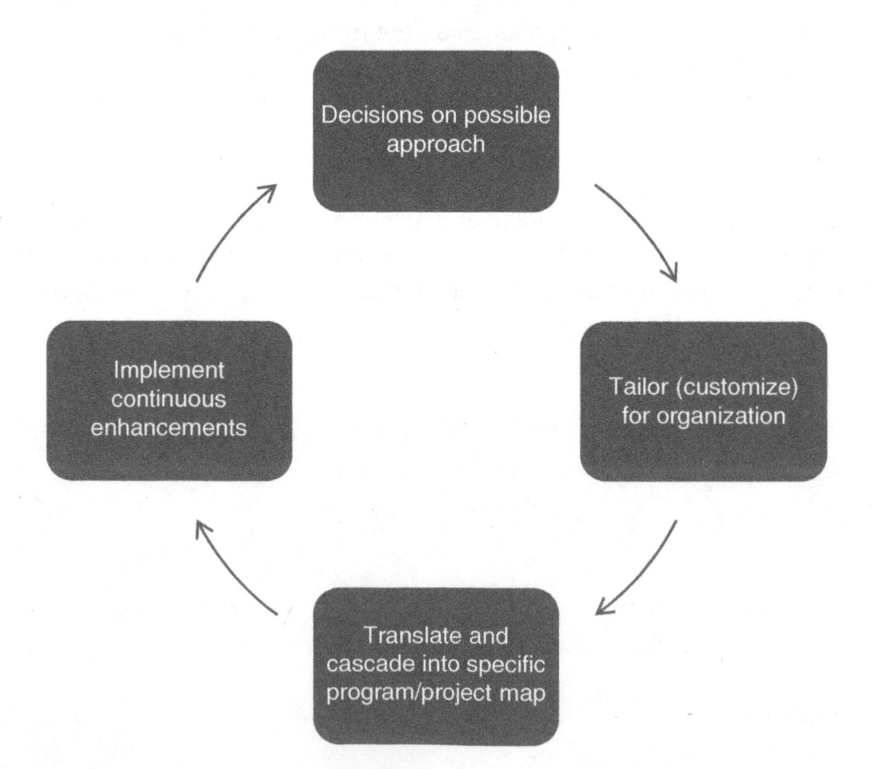

FIGURE 1.6 Tailoring Concept
Source: Adapted from PMBOK® Guide – Seventh Edition.

- Not a one-size-fit all
- The importance of being sensitive to the needs of the organization and the program, and most critically the benefits expected by the customers/users

> **TIP**
> The delivery approach of a program (and its components) has to fit the business context. The program manager should understand the reasons to use for that tailoring.

1.2.7 Reasons for Tailoring a Program

There is a growing list of reasons why the leader of a program would need to tailor the delivery approach. The following list is a starter kit for what you could think about, yet you should always be encouraged to adjust based on your organizational business context and your own program's ecosystem.

- Program's complexity level
- Program team's skillset
- Most suitable delivery approaches
- Program constraints (what might be limiting your results delivery abilities?)
- Global spread of program team (how do I tailor my way of working to best serve the distribution of teams and the virtual settings?)

Review Questions

Parentheses ( ) are used for Multiple Choice, when one answer is correct. Brackets [ ] are used for Multiple Answer, when many answers are correct.

Which of the following is a potential obstacle to achieving programs benefits?

( ) Clear definition of program success.
( ) Extreme focus on program deliverables.
( ) Good change-management process.
( ) A comprehensive benefits-management plan.

How does EFQM align the connected view of portfolios, programs, and projects? Choose all that apply.

[ ] Sharpens the role of culture on achieving outcomes.
[ ] Expands the role of data in creating value.
[ ] Supports the change focus of programs.
[ ] Highlights the program manager's leadership qualities.

What is the greatest value of tailoring?

( ) It makes life easy for everyone.
( ) It achieves tighter governance.
( ) It offers the most suitable program delivery approach.
( ) It gives the scrum master choices.

1.3 SPEED AND QUALITY OF DECISIONS

Decision-making is a key ingredient of program's delivery success. There are many reasons why decisions could end up being inefficient or ineffective. Speed requires autonomy, alignment, and continuous access to the right data that supports making those program decisions timely.

Making the right decision, after having weighted the alternatives, requires leadership, holistic picture understanding, and balancing of the multiple competing interests and possibly complex constraints. Reaching high-quality decisions level is an ongoing development of a muscle many organizations struggle with.

Key Learnings

- Review Programs constraints and stakeholders competing demands
- Using the Samsung case to demonstrate alignment (how the organization demonstrated that)
- Exploring an Innovation model in delivering programs and projects (enhancing the quality of program success)
- Using the ABCD tool of Estimating as an example to address decisions biases (enhance the ability to make better decisions)

1.3.1 Programs Constraints and Stakeholders

Stakeholders could be grouped into the buckets of being interested, involved, or impacted, or a combination thereof. Understanding where the program stakeholders' stand and having a plan to manage the engagement with them minimizes the wrong timing of unpleasant surprises. Having the right balance between program constraints and the positions of its stakeholders supports a program's success.

- Establishes minimum acceptable criteria for success
- Defines methods by which criteria will be measured and communicated
- Ensures consistency with expectations and needs of key stakeholders (within the competing demands)
- Reinforces program alignment for maximum benefit delivery (across the needs of the core stakeholders)

1.3.2 The Samsung Case

Enhancing the organizational decision-making ability is directly linked to the level of alignment across its stakeholders. Figure 1.7 highlights an interpretation of what Samsung has potentially changed over the years to enhance the success of the organization in being a leader in innovation while remaining centered on the customers' continually changing needs.

- Create the creativity and focus on competing stakeholders' demands
- Right balance of governance to continue to achieve speed
- Safety in the exchanges
- Key example to maintain the innovation focus

TIP

Programs, when linked to business value, and are executed as part of an innovation strategic priority, result in stronger alignment and more effective decisions.

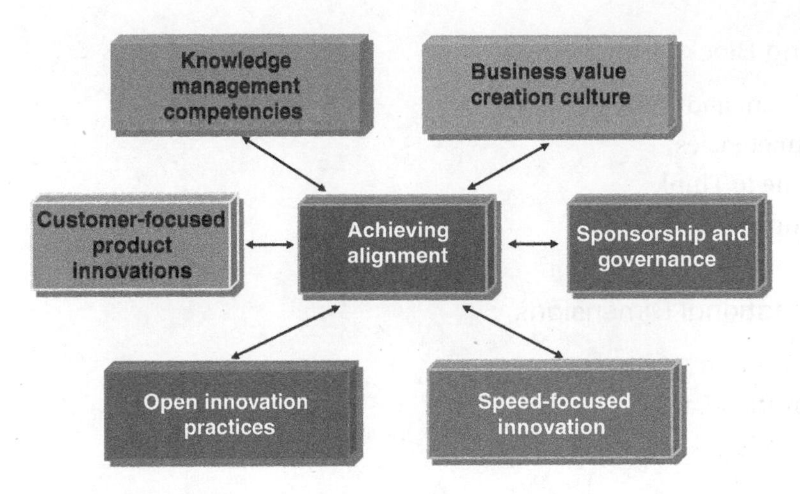

FIGURE 1.7 The Samsung Alignment Sample

1.3.3 An Innovation Model

Innovation models or frameworks help connect an organization and its product and program teams. In Figure 1.8, the innovation model demonstrates the importance of why culture, strategy, and ways of working should come together, to give experts and other supporting teams the opportunity to make innovation a priority. This sets the tone for the collaborative environment where quality decisions thrive and program teams enjoy the change outcomes created by their program work.

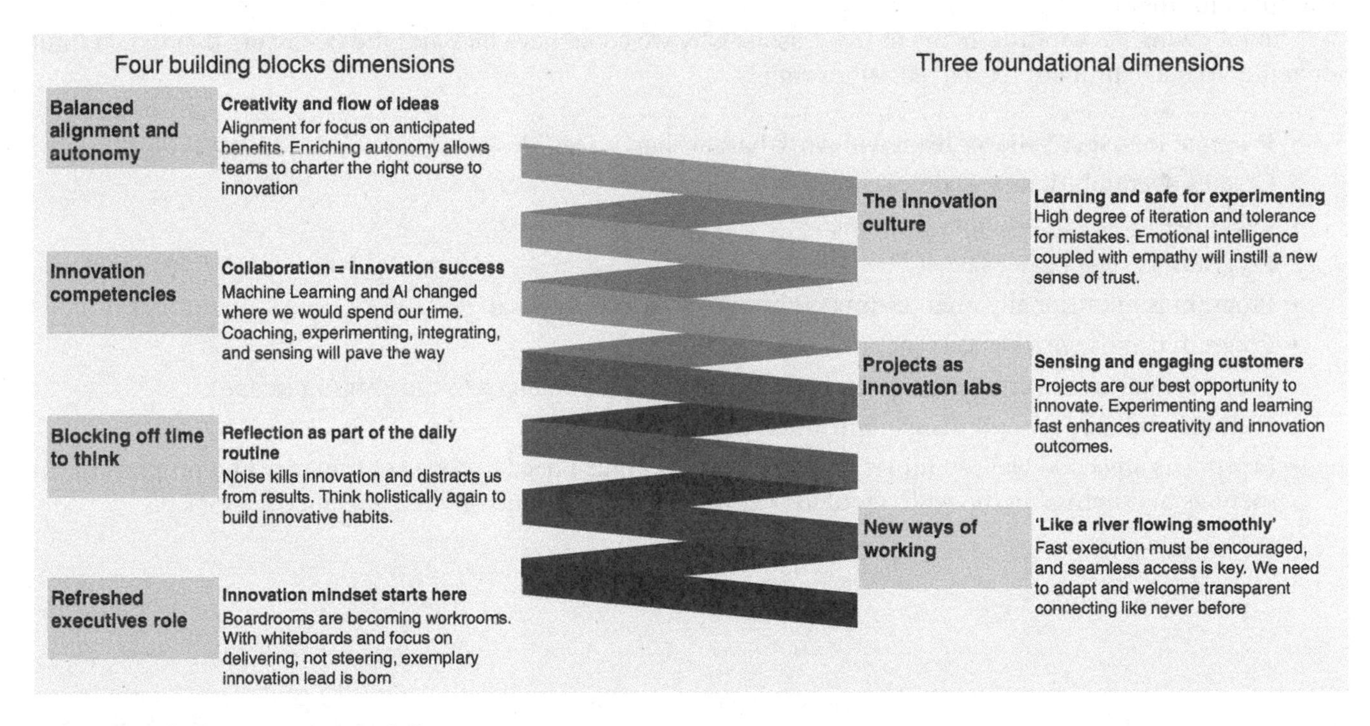

FIGURE 1.8 The Innovation Model[2]

[2]The Innovative Culture Model has been developed by Zeitoun and published as part of the Siemens Digital Industries Software white paper, Simcenter: the heartbeat of the digital twin, adopting a digital mindset to deliver and scale future innovations ©2021 Zeitoun. All rights reserved.

1.3.3.1 Four Building Blocks Dimensions

1. Balanced Alignment and Autonomy
2. Innovation Competencies
3. Blocking Off Time to Think
4. Refreshed Executives' Role

1.3.3.2 Three Foundational Dimensions

1. The Innovation Culture
2. Projects as Innovation Labs
3. New Ways of Working

1.3.4 Why Do We Have Program Estimating Bias?

When estimating the efforts required leading to making critical decisions, care should be taken, given the potential high number of stakeholders and its contribution to creating bias. Bias could come from a variety of sources and unaligned stakeholders' agendas don't contribute to the most effective decisions. A program's wrong or poorly communicated assumptions limit our ability to make high-quality decisions. Tools such as ABCD estimating could help the program manager.

As in Figure 1.9, programs could benefit from the elements of that tool are important to estimating that encompass ways of estimating together with the program team's need to clearly capturing and communicating assumptions, together with the contingencies that reflect an understanding or the program's risks and how to better plan for them.

The following list captures many of the reasons why we could have bias and the behaviors that would limit our ability to make proper program decisions timely.

- Program manager believes team will work harder than is feasible
- Creates overambitious schedules
- Executives want a particular deadline
- Program manager can't say NO
- Program is intentionally underestimated by sales to win a proposal
- Program begins with a good schedule
- Scope creep leads overly optimistic schedule (uncontrolled changed to the program scope)
- Human tendency for optimism (contributes directly to bias)
- Murphy is alive and well, so things will go wrong and thus I need to have contingency and properly document as highlighted in the ABCD tool to positively enhance the quality of program decisions

ABCD estimating

A = Assumptions

B = Basis: expert judgment, analogous, parametric, and, three point

C = Contingency

D = Documented

FIGURE 1.9 Minimizing Estimating Bias

Review Questions

Parentheses ( ) are used for Multiple Choice, when one answer is correct. Brackets [ ] are used for Multiple Answer, when many answers are correct.

When should program metrics be established?

( ) As more program components get executed.
( ) During the chartering process.
( ) As the stakeholders deem it necessary.
( ) A comprehensive set of metrics should be defined at program closing.

How does the innovation model described support the delivery of successful program outcomes? Choose all that apply.

[ ] Making reflection a part of the program manger's daily routine.
[ ] Expanding the quality thresholds.
[ ] Enhancing balanced collaboration and digitalization.
[ ] Refreshing executives' roles

Which of the four elements of the estimating model shown need to be continuously updated throughout the program lifecycle?

( ) Basis
( ) Documented
( ) Assumptions
( ) Contingency

1.4 THE CONDUCTOR

Describing the role of the program manager could take many forms. One of the favorites could be the orchestra conductor. The anticipated integration emphasis played in the role pf the program manager and the need to align across a diverse set of stakeholders make the conductor analogy a fitting designation. One of the experiences worth discovering though would be to get the opportunity to go deep in understanding the role of the orchestra conductor and see that although it is such a critical role, the quality of the outcome of that musical delivery ultimately rests with the orchestra itself, its training, and its achieved harmony.

Key Learnings

- Positioning the capabilities of the Program Manager with an eye on the conductor role
- Applying the balcony versus dance floor model to create impactful holistic views
- Tool that could be used in multiple settings (being holistic and integrative)
- The Program Charter as the key to the conductor's toolbox
- Why is this an ingredient for tools you could put into action?

1.4.1 What Does a Successful Program Manager Look Like?

One of the team building exercises that are worth implementing would be to get a white board and draw a picture of what a successful program manager looks like! It is always fascinating to see how many different views could result from such a creative, yet simple exercise. In a practical sense, answering this question is an important task to accomplish.

Ultimately a successful program manager achieves the anticipated results of the program, builds a strong team that could repeat these successful outcomes in other future programs, and is a leader who is able to properly balance, people, process, and technology as essential for achieving change in the future of work.

The following list highlights what good looks like:

- Aligns the investment in the program to the strategic objectives of the organization
- Strategically aligned
  - Connects the components of the program in a way that increases likelihood of higher efficiencies
  - Utilizes resources well
- Align around benefits
  - Keeps the eyes of the program team members focused on the realization of program benefits
- Stakeholders can be all over the place so I need to use my conductor value to bring alignment back
  - Integrating across program stakeholders like a conductor would
  - Connect the dots
- Make sure the orchestra is delivering

> **TIP**
>
> Your success as a program manager is linked with how well you practice the conductor role to align your team and stakeholders around delivering change outcomes.

1.4.2 The Program Manager Conductor

- As clearly shown in Figure 1.10, there are strong analogies between playing music and running a program. The role of a Conductor is to unify a large group of musicians into a core sound instead of a wild bunch of different sounds surging out
- The program becomes similar to the nicely played piece like the one we enjoy going to the theatre for
- Sees what good looks like
- Steps back and sees the cross dynamics
- Brings the team toward benefits and strategic outcomes

1.4.3 What Could Happen If We Don't Look Holistically?

This conductor analogy continues to help us rethink the role of the program managers. Since strategy is about clear purpose and its execution is about getting resources aligned and moving toward that purpose, this analogy works. The ability to be that holistic and see how the parts are continually connected is such a critical program muscle.

The following could be consequences when we don't maintain that holistic view:

- Losing sight of what strategically matters
  - This is naturally a disaster
  - Missing the right priorities

FIGURE 1.10 The Conductor
Source: takazart / 360 images.

- Lacking clarity in programs' roles and responsibilities
  - Many data points confirm that this contributes to program failure
  - You want to reverse that
- Achieving deliverables' delivery success and failing to get customers what they truly need
  - Been in so many cases where a client could be in a different place than my assumption that we are ready
- Program team's burnout and mental fatigue and associated losses in productivity and creativity
  - Program manager should maintain the clarity and objectivity of views

1.4.4 The Analogy of Balcony Versus Dance Floor

Another analogy, as highlighted in figure 1.11 relates to the importance of the strategic and holistic view is being on the balcony versus being on the dance floor. Leaders realize the importance of this. This contributes to the adaptable mindset that program managers should possess. Program managers tend to be more successful when they have the ability to create the distance and see better where some of the gaps might be in what is happening in front of their eyes on the dance floor, or in the deep work the program team is involved with. They should also have the ability to roll up the sleeves and jump right back into it and be in the trenches with their program team colleagues.

A few points to remember about this tool:

- A tool that goes directly to the toolbox
- Creating the distance
- Action is great, yet you got to find the moment to step away
- Create the objectivity and holistic view
- See the politics and the gaps of what is happening or not

FIGURE 1.11 The Holistic View
Source: StockSnap / 27551 images.

1.4.5 The Program Charter

There is an increasing programs complexity. Technology demands are increasing, cyber security is a major issue, sustainability is a major strategic priority, and stakeholders' expectations are steadily increasing. These continual competing demands on the program manager require that a tool for laser focus is used. Without such focus, the likelihood of expectations creeps (e.g. why don't you also do this while you are at it?) and will also continue to grow and programs and projects will not meet their expected destination targets. When this is coupled with ownership gaps (e.g. it's not my job), then alignment matters most.

The Program Charter (the best tool in our arsenal for common vision).

When we look at the charter for its true value given that it provides the guardrail for why we do the program/project and the expected view of what success looks like, we see that it is also a great opportunity to say NO. Saying NO is a muscle that has to be well developed for the program managers.

Among the most valuable reasons to have a program charter are its ability to address:

- Clear view of program success/associated strategic benefits (Without it, no clear view of what success is to be achieved)
- Strategic metrics/guideposts to confirm eventual acceptance (A few short concise pages that give us a strong staring point)
- Strategic risks/assumptions to consider (at least high level)
- Program governance framework (key abilities to make decisions, within and outside your control and the close relationship to the program sponsor)

1.5 ELABORATING THROUGH COMPLEXITY

One of the determinants of the future of work is the ability to adapt. Programs' lifecycle typically encounter multiple changes. The ability to be resilient in the face of the changes and to act decisively and quickly to adjust, is becoming one of the top ranked qualities that program managers should possess. Elaborating through complexity means that we start with what we know and continue to peel the onion, get more information, ask the right questions, and exercise immense amount of collaboration within and across teams.

Key Learnings

- Progressive Elaboration as a theme: "Ongoing dynamic view of taking complexity into something simpler to handle"
- Balancing uncertainty, change, and complexity
  - Unique view of managing program success
  - Different from the land of projects
- Using the Scaled Agile Framework (SAFe) model to manage scaling complexity
  - Great opportunity to look at different ways to scale complex programs
- Highlighting clarity and cascading complexity using the Facebook case
- Identifying the role of continual prioritization and progressive elaboration in driving programs success
  - Key mechanism for the way to plan and execute the work of the program

1.5.1 Balancing True Program Constraints

In the larger and more integrated setting of a program, it is no longer your classic project constraints of cost, schedule, and scope. Figure 1.12 shows program constraints expand to cover complexity, uncertainty, and change. This is going back to the strategic role programs play. This shift from the traditional scope, cost, and schedule constraints is critical in order to balance the true pressures placed on the importance of achieving program benefits. Many programs could easily fail if they focus on the classic constraints and end up winning the battle and losing the war.

Central focus on benefits (understanding the role of aligning the program team and the stakeholders strategically around the benefits) becomes critical. This has to be coupled with high degree of sensitivity to the changing work dynamics (mix of the team, ways we govern, competing demands, all have to be balanced).

TIP

Remember programs are unique in their focus and conditions. Change your perspective from the classical project constraints to the program ones.

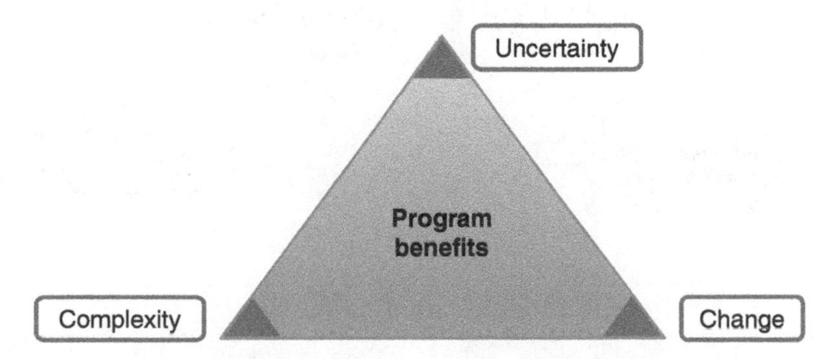

FIGURE 1.12 A Program's Competing Constraints

1.5.2 Scaled Agile Framework (SAFe)

The framework is a useful instrument for scaling complexity in programs and across the organization. Given that SAFe could look across the enterprise and tackle the required governance across the layers of portfolio, large solution, program, and then further down to the team, cascading and mapping across these layers work well. In the program layer, the ability to have the focused release train concept drive experimentation, integration, deployment, and on demand release enhances the opportunity of effective delivery of programs' value. Connecting metrics, roadmap, and the vision is also instrumental for enhancing the balance through complexity.

Program managers should explore and experiment with the SAFe layers. As a framework, it is helpful in better understand the interoperability between portfolio, program, and projects. To reflect on how deep team interactions take place, one should get into the team layer and reflect on how governance is conducted. Another beneficial aspect is that this framework builds on the Agile practices (incremental ways of delivery leading to a changed way of governing).

SAFe Principles[3] help us manage program complexity:

(1) Take an economic view
(2) Apply systems thinking (critical to the agile mindset and the end-to-end view)
(3) Assume variability; preserve options
(4) Build incrementally with fast, integrated learning cycles
(5) Base milestones on objective evaluation of working systems
(6) Visualize and limit WIP (work in progress/process)
(7) Apply cadence, synchronize with cross-domain planning (e.g. releases and associated iterations)
(8) Unlock the intrinsic motivation of knowledge workers (team layer is well addressed)
(9) Decentralize decision-making
(10) Organize around value (critical outcome from the framework)

1.5.3 The Facebook Case

Leonardo da Vinci said: "Simplicity is the ultimate sophistication." In the case of programs and program management, he would have not been more correct. With the massive data, number of users relationships, partnerships, and many other resources to balance, as reflected by figure 1.13 in the case of Facebook, it is critical that we find

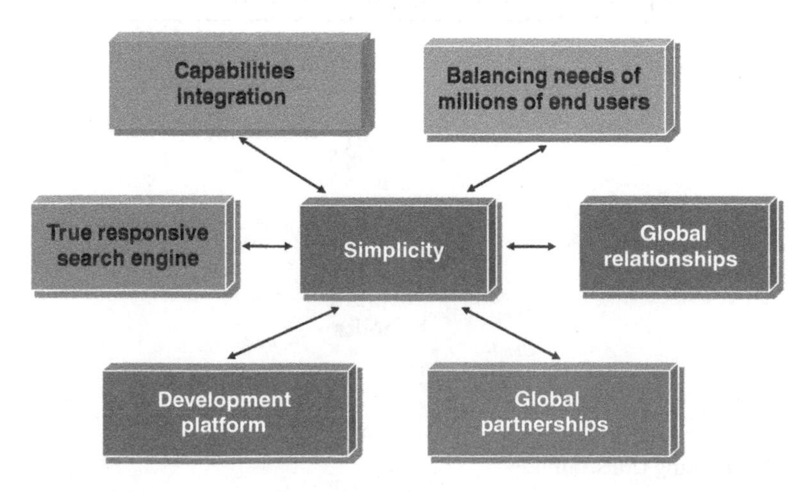

FIGURE 1.13 Facebook and Simplicity

[3]https://v5.scaledagileframework.com/safe-lean-agile-principles.

the recipe for what works for a given program or a portfolio of programs and projects that would allow us to achieve simplicity, create clarity, and communicate in a consistent way across the organization and the program team.

- Simplicity is the core in order to handle program complexity
- Global mindset in relationships and partnerships building and investing
- Create responsiveness that aligns with the behaviors and patterns of different program users
- Handling the complexity for such a mega community of users is a strong example of building simplicity into a program's way of working

1.5.4 Progressive Elaboration

This concept is critical for the future of programs and projects success. The key takeaway of using this tool is continuously elaborating as more details become available. This helps us deal with complexity, allows us to make progress, and ensures getting value out of the door and in the hands of customers and users faster. This is also why this concept or tool is an important principle in portfolio, program, and project management certifications.

The idea is that the program managers start with what they know and then roll with it as they uncover more meaningful details and get wider access to relevant data. This ability to adjust and shift allows the program manager to also play a key role in driving initiatives that relate to Environmental, Social, and Governance (*ESG*) strategic objectives that organizations prioritize.

To summarize the most useful dimensions of this concept:

- Usually refers to ability to add more details as uncertainty decreases
- Critical way of thinking and working in programs and projects today given the need for speed and the frequent occurrence of starting programs without full understanding of requirements (balance between what it minimally required and what I can extract later on)
- Supports organizational agility principles (helps on both waterfall and agile delivery models)
- Helps us with customization of best fitting program delivery approach

The value of the ***progressive elaboration*** way of working is further expanded when we consider the elements of the VUCA conditions that surround most modern programs. This requires the program manger to adapt well to handle the amount of volatility, uncertainty, complexity, and ambiguity that is increasingly the norm for running and changing the business. Additionally, since each project and a program is surrounded by Enterprise Environmental Factors (*EEFs*), many of which are outside the control of the program manager, there is a growing and vital role for progressive elaboration to address these program constraints.

Review Questions

Parentheses ( ) are used for Multiple Choice, when one answer is correct. Brackets [ ] are used for Multiple Answer, when many answers are correct.

How does the analogy of conductor relate to the role of the program manager?

( ) Programs are like music.
( ) It illustrates the need for strong control of team members.
( ) It shows the importance of the alignment of various team members' roles.
( ) It is a way to show the team who is boss.

(continued)

How is having a holistic lens critical for programs success?
Choose all that apply.

[ ] Reflecting on how the pieces are fitting together.
[ ] Getting onto the balcony is always great.
[ ] Ensuring that the program manager is not involved in the details.
[ ] Enhancing the leader's objectivity.

Which of the four elements below is valuable to include in the program charter?

( ) Detailed deliverables
( ) Documents control process
( ) Success definition
( ) List of all stakeholders

When is progressive elaboration most valuable?

( ) As program managers become more experienced.
( ) In most cases, because there is usually a level of uncertainty in programs that would only unfold over time.
( ) When the number of stakeholders increases.
( ) When clear and comprehensive requirements are in place

How does SAFe help us in handling and scaling complexity?
Choose all that apply.

[ ] Build incrementally with fast, integrated learning cycles.
[ ] Base milestones on meeting delivery deadlines.
[ ] Centralize decision-making.
[ ] Refreshed executives' roles.

Which of the following elements contribute to Facebook's illustration of handling complexity?

( ) Addressing the needs of select end users.
( ) Increasing documentation.
( ) Integration capabilities.
( ) Staying focused on the Facebook way of working.

1.6 MANAGING CHANGE MATTERS

This part of the chapter is focused on programs' contribution to transformation and achieving change.

1. Understand the balance between running the business and changing the business.
2. Reflect on change models that help you develop your change agility.
3. Learn the critical ingredients for successful digital transformation.
4. Explore the change maker role and its value across the program lifecycle.
5. Develop an appreciation for the engagement strategies necessary for creating buy-in.

Key Learnings

- Important foundation for shifting to the change management mindset
- Using the SAP story to highlight a critical solutions implementation success enabler
  - As an example of a mega enterprise solutions organization to show the important of the change management mindset
- Defining the meaning and connection between run the business and change the business
  - Separating the project and program way of doing things from the traditional getting stuff done
  - Why does creating this understanding matter for the success of program work?
  - Culture changes in the organization
- Connecting change to achieving strategic outcomes
  - Very important element to the end-to-end connection needed for program work

1.6.1 The SAP Story and Achieving Change

The theme for this story is to take an example of a high tech global organization and exploring the following focus: Imbedding the change management practices into the implementation rollout.

- Early realization of the criticality of change management in the successful roll out of enterprise solutions worldwide (Managing resistance is critical to managing changes in the created new ways of working.)
- The importance of achieving a common level of program/project management maturity across the organization (in this case by looking holistically across regions and using some model like Kerner's or Capability Maturity Model Integration (CMMI) to achieve maturity commonality in use of change and project/program management practices)
- Linking the performance of program managers to the achievement of the desired program change for customers (what we don't measure does not create the right supporting behaviors)
- The developing role of the Global PMO (to help connect ownership like in this case of working across a global organization, centralizing ownership supported the repeatability of certain success patterns)

1.6.2 Run the Business and Change the Business

As much as running the business is critical for sustaining the organizational footprint, it is changing the business where the innovations are and the growth excitement resides. Programs and projects are the vehicles for achieving such change. Figure 1.14 highlights that a balance is needed, while we realize that as we continue

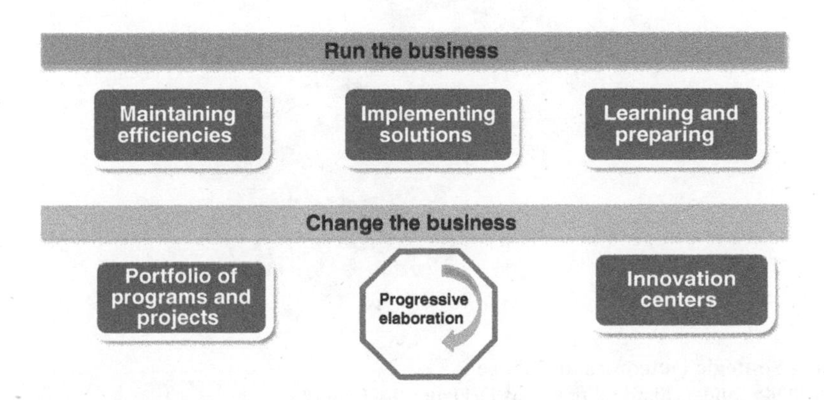

FIGURE 1.14 Achieving Change Balance

moving toward the program/project economy of the future, the balance will be tilting toward changing the business, and digitization could likely take care of most of the running the business parts of the future organization. This will continue to have distinct implications on the make-up of program teams and the relevancy that program team members and experts bring to the mix.

- Running the business is done by many, as in the case of business units and back-office entities
- Changing is not about just maintaining efficiencies, or learning for the next time around
- Changing the business requires a different set of skills, e.g. the 3Ps or portfolio, program, and project cascade or the capability to progressively elaborate and continually refine and iterate
- Changing motivates the innovative mindset as highlighted by the innovation model in earlier lesson to bring continual innovation and creative ideas that support excellent practices in the change environment associated with program work

1.6.3 Connecting Program Change to Strategic Outcomes

This cascading point is important in reminding us that programs create change that is connected to strategic outcomes. This cascade ultimately connects to operations, where running the business is. This also means that there is continual feedback loop that affects the future mix of programs and projects and the next set of portfolio choices.

Programs are focused on creating the valuable change to organizations

- This view in Figure 1.15 confirms strategy drives portfolios that are broken down into programs and projects. The outcomes go into operations and continuous feedback affects the prioritization and future changes
- That connectedness owned by a PMO, a sponsor, or other leaders is a critical to the change focus

The dynamic nature of feedback and continual learning contribute to realizing change value:

- Openness is key ingredient to program work
- Knowledge management focus is central to this process
- Continual success

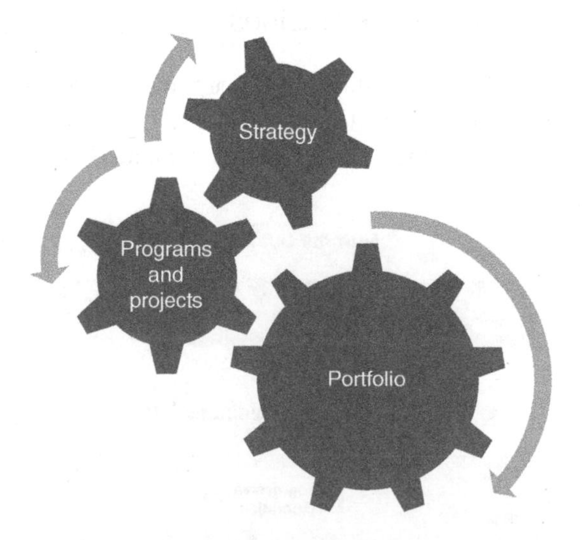

FIGURE 1.15 Connecting Strategic Outcomes to Change
Source: Adapted from the PMBOK® Guide – Sixth Edition (Part 1), Figure 1-4, Page 17.

Review Questions

Parentheses ( ) are used for Multiple Choice, when one answer is correct. Brackets [ ] are used for Multiple Answer, when many answers are correct.

Making sure that there is a continuous revisit back to the strategy is fundamental to programs. Why was change management highlighted as a critical muscle in the SAP story?

( ) Technical details of the enterprise solution
( ) In most cases, there could be a perception of complexity and resistance to implementing new solutions.
( ) When the customer asked for it.
( ) Too many chefs in the kitchen.

What has been driving the increasing focus on growing the business?
Choose all that apply.

[ ] The Project Economy is here.
[ ] Executive maturity in understanding the role of programs/projects.
[ ] Boredom with operations.
[ ] The fun of growing the business.

Which of the following illustrates the importance of connecting program changes to strategic objectives?

( ) Change is difficult.
( ) Increasing governance.
( ) Achieving the change value.
( ) Staying focused on the bottom line.

1.7 ALIGNING ACROSS HEARTS AND MINDS

Key Learnings

- A word that program managers continue to think about and use is alignment
- We look at models and agility topics to align the hearts and the minds
- A real cascade to create focus is central to this lesson
- The worst scenario – not being aligned
  - From real-life stories, this could be a nightmare as it shows major program gaps such as in the case of vague roles and
- Linkages across phases and enablers of change in ADKAR's Model
- Use Kotter's 8 steps to drive from creating urgency to embedding change
  - Both models create energy for change, its urgency, and what it takes to implement change
- Examples of change agility
  - Why is this becoming a very important quality that future leaders globally need to possess in the next decade?
  - Builds on the changing the business aspects from previous lessons

1.7.1 The Case for Alignment

Alignment is one of these power skills of program managers. This contributes to multiple opportunities for the program managers to successfully move the needle closer to achieving the strategic value of the programs work. Here is a possible list of what alignment contributes to:

- Improving performance; by linking to the strategic direction of the organization
- Empowering people; by allowing them to self-organize when there is enough alignment balance
- Motivation
  - Understanding how the pieces fit
- Allowing experimentation; given the availability of a guiderail
  - Key ingredient of future organizations
- Improving program teamwork; with clarity of roles and responsibilities
  - Reaching a high performing team
  - Trust becomes the foundation
- Motivating stakeholders; given their ability to engage with clear programs' direction
  - Motivating for the extended program stakeholders as they see and experience how well the program team is connected

1.7.2 ADKAR Model[4]

- Figure 1.16 highlights ADKAR's enablers. Key elements around awareness, desire, knowledge, and reinforcement

Employee phases of change

		Enablers
A	Awareness of the need for change	• Management communications • Customer input • Marketplace changes • Ready-access to information
D	Desire to participate and support the change	• Fear of job loss • Discontent with the current state • Imminent negative consequence • Enhanced job security • Affiliation and sense of belonging • Career advancement • Status or change in social standing • Acquisition of power or position • Incentives or compensation • Trust and respect for leadership • Hope in the future state
K	Knowledge on how to change	• Training and education • Inform action access • Examples • Role models
A	Ability to implement required skills and behaviors	• Practice applying new skills or using new processes and tools • Coaching • Mentoring • Removal of barriers
R	Reinforcement to sustain the change	• Incentives and rewards • Compensation changes • Celebrations • Personal recognition

FIGURE 1.16 ADKAR Model and Enabling Change
Source: Adapted from https://www.prosci.com/methodology/adkar.

[4]https://www.prosci.com/methodology/adkar

- Linking programs to purpose
- Designing program lifecycle phases with change management in mind (e.g. the SAP change story and using the model to remind us of key enablers, like training, taking fear out of the way, or running meetings differently)
- Using ADKAR to understand the focus required for realizing benefits (review each of the elements in relationship to realizing benefits)

1.7.3 Kotter's 8 to Focus on Implementing Change

Figure 1.17 shows a simplified view of Kotter's 8 model. This is an easier model to relate to and implement from a program standpoint. As many programs are executed to drive transformational change, the 8 steps in this model map nicely to what program managers would need to do to properly plan and to prepare the organization for the anticipated change. The 3 buckets of creating the climate, engaging, and then implementing the change map nicely to the program lifecycle and the increasing focus on achieving strategic benefits.

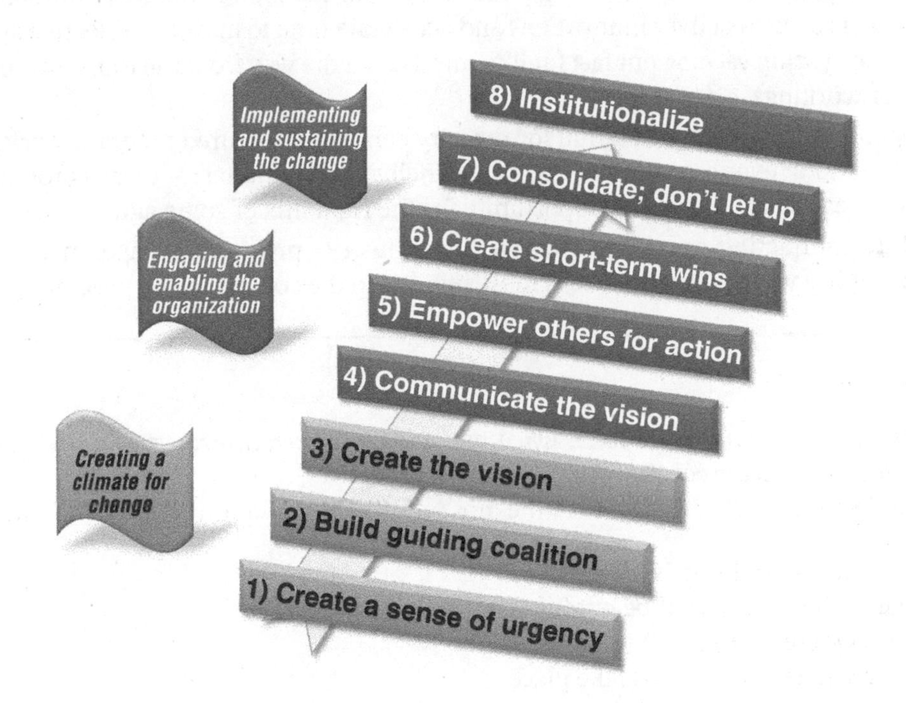

FIGURE 1.17 Kotter's 8 and Implementing Change
Source: Adapted from https://www.kotterinc.com/methodology/8-steps/.

1.7.4 Creating a Climate for Change

Create a Sense of Urgency – changing the business is what we are after in program work, focus
Build Guiding Coalition – across stakeholders and stakeholders' groups
Create the Vision – ultimately achieve a climate that ensures the vision is clearly understood

1.7.4.1 Engaging and Enabling the Organization

Communicate the Vision – simplicity, like in the Facebook Story, or the cascading using the progressive elaboration
Empower Others for Action – commitment matters most to creating the change
Create Short-Term Wins – agile or not, try to find ways to create the wins fast and establish commitment

1.7.4.2 Implementing and Sustaining the Change

Consolidate; Don't Let Up – momentum around the change (key to the role of the program manager in connecting the dots)
Institutionalize – sustaining the change created by the program's benefits

1.7.4.3 Change Agility Examples

Builds nicely on instituting the change

World events, COVID, and disasters lead us to expect that things are done in a difference ways and cadence. The following reflects such a climate for change:

- Organizations applying a varied mix of waterfall and agile across business lines
  - The future of work is hybrid
  - A pharma company having to follow rigorous process for the high regulatory compliance components, whereas agile is used to deliver innovations and accelerate time to market – selective application of agile principles, e.g. getting vaccine out fast (agility mindset all the way from the top, culture adaptability, and new ways of working)
  - An energy company applying waterfall for the very complex structured program work, while its IT uses agile practices to deliver a continuous stream of deliverables and fast – there is room for both process repeatability yet you can imbed an opportunity for the right mix of some agile principles
- Agile models of contracting even in the construction industry – program manager makes the right decision of how much agility versus set structure is most suitable to the context of the program

Review Questions

Parentheses ( ) are used for Multiple Choice, when one answer is correct. Brackets [ ] are used for Multiple Answer, when many answers are correct.

In the ADKAR's model, where do you see the most emphasis on aligning hearts and minds?

( ) In the "K" phase where training is provided.
( ) In the "D" phase where a sense of belonging is created.
( ) In the "A" phase where barriers are removed.
( ) In the "R" phase where celebrations take place.

In which of the following elements of Kotter's model do you see the alignment with the program stakeholders' engagement necessary for implementing change?
Choose all that apply.

[ ] Build guiding coalition.
[ ] Create short-term wins.
[ ] Institutionalize.
[ ] Communicate the vision.

Which of the following shows how the use of a mix of program frameworks supports change agility?

( ) Using a consistent method to address the needs of the customer.
( ) Total autonomy.
( ) Applying waterfall for complex program work, while agile practices help deliver a continuous stream of deliverables.
( ) Staying focused on increasing alignment.

1.8 DIGITAL TRANSFORMATION

Key Learnings

- Organizations into the next decade will continue to think digital transformation, technology reliance is increasing, and the pace is getting faster
- This lesson gets into the transformation mindset and culture, and it is a chance for program managers to show their caliber in working with stakeholders
- Humans 2.0 or 3.0 are the future of getting the right balance between humans and technology
- Need to think of the elements that could be applied on a daily basis to ready the teams and organizations for digital transformation
- The myth behind digital transformation and why 70% of these initiatives fail
  - Data points from Gartner and others that confirm a consistent high failure rate
- The role of culture in the necessary programs' experimentation safety
- Explore the connection between proper program planning and being in the 30% of transformations that succeed
- Resilience – The Antidote for Change
  - Why is that quality such an important component to inject in the digital transformation programs to increase their chances of success?

1.8.1 Why Do 70% of Transformation Programs Fail?

- It's not the technology mix
- It's not just the lack of proper planning (sometimes there is even a massive investment in planning and even great interest from executives)
- It is the gaps in designing the purposeful ingredients:
  - A collaborative culture
  - A focus on guiding principles
  - The human enablement (the future for the program managers is to ensure that we are highly centered on the human element)
  - The right level of digital integration (make sure that the right level of integration coupled with the proper assumptions)

TIP

The human element is central to the success of transformation programs. These programs don't fail because of technology.

1.8.2 Digital and Programs Decision-Making

The future of work has vast opportunities for the program manager with a variety of decision-making tools. This is going to be critical in rethinking where do program managers spend their time and create the most value of their roles.

FIGURE 1.18 Data-Driven Decision-Making
Source: geralt / 25597 images.

As illustrated by figure 1.18, digitalization will continue to enhance the speed and quality of decision-making:

- Digitalized economy – capitalize on intelligence
- Artificial intelligence
- Internet of things (IOT)
- Use of "Big Data" – find ways as a program leader to focus much more on trends, program direction, and doing half time adjustments, thus directly affecting program success
- Analytical statistics

1.8.3 Culture and Programs' Experimentation Safety

Program managers should have a very good sense of their environment and realize the criticality of having the right supporting program culture. There is a strong need to rethink things and to continuously remember that culture is fundamental to the program's success.

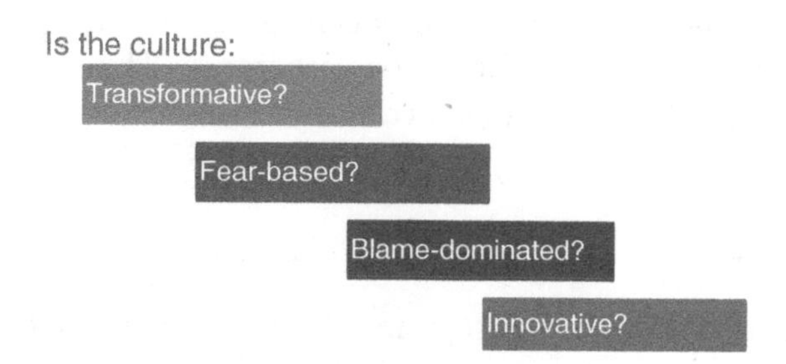

FIGURE 1.19 What Is the Dominating Organizational Culture?

As captured in figure 1.19, program managers and change leaders should make sure:

- It is transformative
- It is not fear infested
- That you get rid of blame dominance as it directly negatively affects success
- Transformation success hinges on how innovative the team is

1.8.3.1 Culture and Programs' Experimentation Safety

- Important in operating with agility and equally in digital transformation
- Cultures that focus mainly on command and control would not be effective in the future
- While the program manager could create an experimentation program team culture, it needs to be supported by a safe corporate culture that encourages ideas to flow freely, is transparent, and rewards collaborative behaviors including openness of feedback
- This creates more fun and innovation and contributing one-2-one to the success of digital transformation programs' success

1.8.4 Effective Program Planning and Digital Transformation Success

As highlighted earlier in studying the reasons why digital transformations fail, effective program planning is a must if we would want to effectively turn the ship that is headed for disaster and to safeguard organizational resources.

- Link to strategic planning
- Doing the right steps upfront as in the program charter
- Formulating the roadmap

Understanding the environment especially with the increasing volatility around us, this leads to a sold program plan

- Ensuring the alignment to a clear strategic purpose
- Rigorous chartering and sponsoring process that considers risk and the environment – creating a higher and closer understanding of the risk, thus the proactivity needed for digital transformation
- Builds on a clear roadmap that strongly connects stakeholders – envisioned pathway from A to Z
- Needs to address digital transformation with the multitude of change management skills we have been addressing across this chapter

1.8.4.1 Resilience

Resilience is a key contributor to better handle change and directly contribute to programs' success:

- Part of the nature of change
- Seen in the dynamics around programs
- To handle dysfunction in the program team
- To address lack of clarity of success upfront
- Needed to polish one's leadership quality
- Curiosity muscles development needed to ensure how ready we are
- The capacity for a program team to absorb high levels of change while displaying minimal dysfunction
- Sets the stage for a new definition of program success
- Is not purely a trait, it can be developed and nourished (needs to put myself in complex situations, dealing with politics and ambiguity as part of the nature of the world we live in)
- A leadership competency that ties to the curiosity muscles (role of knowledge and managing knowledge)
- Dominates future workplaces (centered on how resilient could I be)
- Links to the growing project economy focus (a growing future realization in terms of understanding the strategic role of project and programs and thus the need to exhibiting resilience in that future organization)

Review Questions

Parentheses ( ) are used for Multiple Choice, when one answer is correct. Brackets [ ] are used for Multiple Answer, when many answers are correct.

Which of the following contributes the most to successful implementation of digital transformation programs?

( ) Rolling out with a focus on what each functional area needs.
( ) Creating a comprehensive mix of technology solutions.
( ) Detailed planning.
( ) Building a collaborative culture.

Which of the following elements contribute to the healthy culture needed for successful programs outcomes?

Choose all that apply

[ ] Fear
[ ] Experimentation safety
[ ] Command and control
[ ] Free flow of ideas

Why is resilience the trait that is expected to dominate future workplaces?

( ) It is a cool trait.
( ) It is similar to emotional intelligence.
( ) It absorbs high levels of change and creates high performance.
( ) It supports resisting program change.

1.9 THE CHANGE MAKER

Key Learnings

- Make the impact of the program manager complete the notion of being a change maker
- Especially due to the strategic role pf programs
- A must learn set of attributes and qualities that make us stronger in creating change
- As a reminder the balance between running the business and changing the business
- What and how to develop those qualities needed for our role as program change mangers
- The SAP example and how to motivate change makers – making sure implementation is connected with the change
- The learning sponge as a way of leading change – capitalize on that to lead change
- Understand the meaningful program lifecycle points that make change successful
- Map the EFQM model leadership and process elements to the change maker role – how much is there emphasis on the right change attributes

1.9.1 The SAP Story and Change Makers

- Inclusion of change in SAP's solutions implementations and sophistication of roll outs worldwide
- The realization that all solution rollouts are making critical changes in the customers organizations
- The advancing program delivery capabilities that are resonating with change makers and next-generation PMs – how he or she will drive change, what is required, and what is expected of the users in the organization's SAP works with
- The maturing of the view of program management as a different making capability – next-generation program management should understand the change making role
- The strength of program management in connecting the stretching strategic agendas of organizations, including DE&I and ESG – continuing to make a difference in the form of the program work, stretching my capability organizationally speaking in terms of tying the work to the strategy
- Which means that organization executives are attributing their success to it! – getting executives' attention to ensure the importance of the change making function, this will also show that program management has reached an excellence level in the corresponding organizations

1.9.2 The Learning Sponge

- Figure 1.20 reflects some key building blocks to developing the learning sponge such as: The importance of knowledge management
- Sets the foundation for a program manager to stretch the leadership qualities
- Opening the appetite for learning
- Focus on ways by which learning can support the change
- Unlearn, relearn, and learn some new things
- Thus grow the roles and functions of the program manager and spill that learning over to the program team and the surrounding stakeholders' environment

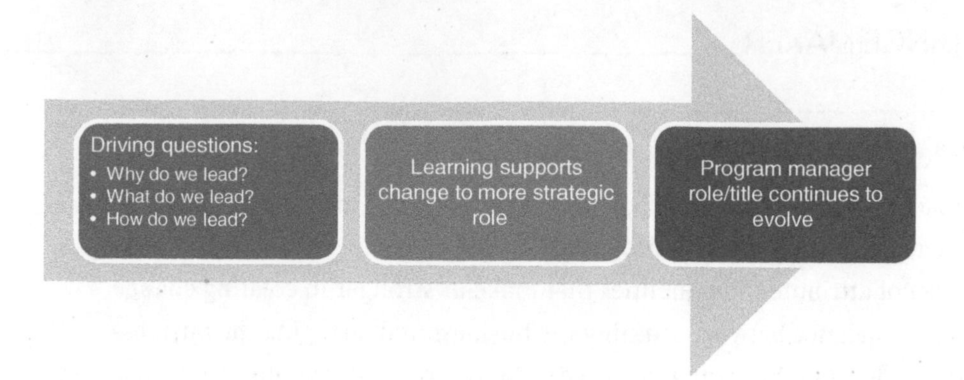

FIGURE 1.20 Developing the Learning Sponge

1.9.3 Program Lifecycle Points that Make Change Successful

As we look across the program lifecycle phases in the Standard for Program Management – Fourth Edition, and see the connections and related detailed steps in the 3 phases of program definition, program delivery, and program closure, one would notice the importance of:

- Connecting change to the program lifecycle
- Activities to plan for, monitor, control, and administer changes during the course of the program
- Nothing should be managed in isolation
- We are unifiers, integrators, able to see holistically, getting on to the balcony
- Supplementing the lifecycle phases with change management elements
- Update the learning along the way as we move toward closure
- Become an organization and team able to administer changes in a way that is clearly aligned with the program journey
- Practice change management throughout

1.9.4 EFQM Model and Change Maker Competencies

- Figure 1.21 highlights many of the attributes that are relevant to the role of the program change maker
- Decision-making and the quality of decisions matter
- Can't afford weaknesses in decision-making muscles

FIGURE 1.21 Change Maker Competencies

- Balancing the people and technology side with having the right processes in place
- Leadership qualities, like relating with compassion while maintaining the value focus and excelling in the communication as the key connector
- Similar to what we described earlier about the value of the conductor role in what a program manage does

Review Questions

Parentheses ( ) are used for Multiple Choice, when one answer is correct. Brackets [ ] are used for Multiple Answer, when many answers are correct.

Where are the key overlaps between change management activities and the program lifecycle?

( ) Program delivery only.
( ) Across all the lifecycle phases.
( ) Program delivery and program closure.
( ) Program definition and program delivery.

What elements of the SAP story showcase the change maker's role?
Choose all that apply.

[ ] The realization that all solutions rollouts are making critical changes in the customer's organizations.
[ ] The maturing of the view of program management as a difference-making capability.
[ ] Change management system.
[ ] Maximum autonomy.

Which of the following is a critical change-making competency?

( ) Timeline focus.
( ) Scope-change management.
( ) Transformation skills.
( ) General management skills.

1.10 CHAMPIONING CHANGE

Key Learnings

- Why is the program championship required?
- How can engagement strategies create program change buy-in?
  - Enable the program team
- Put servant leadership attributes in action to enable program teams
  - Cross-understanding other importance of that to embed in the team and across the program leaders
- Illustrate how the IOT implementation success is championing change-centric programs
  - How does technology help us mature our change muscle?
- Linkages to PMI Pulse of the Profession "Beyond Agility" to highlight critical future change capabilities
  - How to push ourselves past the agility that we need to imbed in our programs and projects

1.10.1 Engagement Strategies for Buy-In

Figure 1.22 summarizes a set of guidelines to help shape the creation of engagement strategies to achieve buy-in. The achievement of program changes highly correlates to the ability to engage the minds and the hearts of program stakeholders. Creativity in designing and applying the right mix of strategies will help increase the speed and chances for achieving buy-in. The following highlights a multitude of the best practices we could apply to achieve this buy-in.

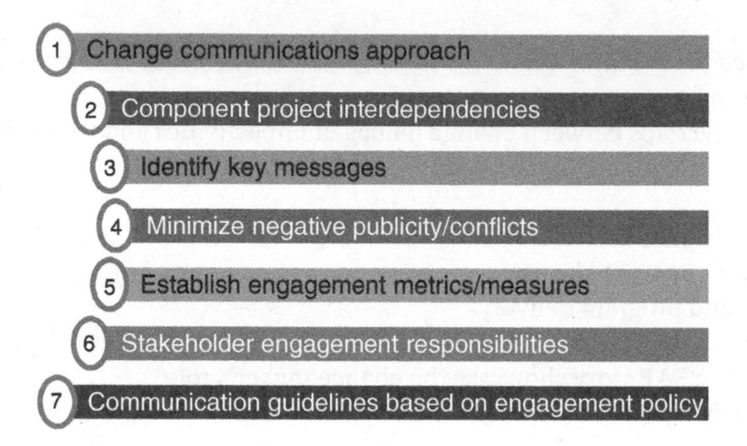

FIGURE 1.22 Engagement Strategies

- Illustrate communication clearly to the team
- Communications strategy and showing how we communicate differently as needed
- Interdependencies illustration as in the roadmap and schedule to show cross impacts
- Repeat the key messages often
- Revisiting the metrics to support taking out the limiting behaviors
- Use value and strategy-based metrics
- Tell me how I am measured, I will tell you how I will perform
- So much need to engage using our change management behaviors
- Clarifying roles and responsibilities
- Thinking what good looks like when it comes to communication

1.10.2 Servant–Leadership Attributes

- As illustrated by figure 1.23, being a servant leader is a huge differentiator for the success of the program team
- Inspires and energizes the team – takes obstacles out of the way
- Utilizes design thinking to empathize – better sensing and understanding of needs
- Addresses cultural transformation – program manager and the sponsor
- Practices emotional intelligence – ensuring maturing the elements
- Participates in change management efforts – rolling up the sleeves, showing by example
- Demonstrates great respect for program team – this motivates across program teams

> **TIP**
>
> The future of program managers success hinges on their ability to be of true service to their teams. Investing in your coaching and connecting skills is essential.

FIGURE 1.23 The Servant Leader as a Difference Maker
Source: sgottschalk / 7 images.

1.10.3 IOT Implementation and Program Change

Given the complexity and the dynamic nature of programs discussed in detail across many of the parts of this chapter, it would be great to continue to capitalize on technology to experiment fast, move and adapt well, and retie the program approach quickly. The Internet of Things (IOT), value of which is highlighted in figure 1.24, is one such key enabler for this enhancing this program change handling capability.

- The IOT and access to data analytics change where we spend our time
- Great starting points to compare issues, risk and pain points
- Simulate and redo things without having massive amount of time and effort investment
- Rethinking using data to come up with better solutions
- When I have the proper nature and amount of data, this can help me have better handle on the program constraints and the elements of VUCA
- Responsive team that are better connected strategically around program risks
- Forecast resources and predict dynamic movements of parts of the programs
- Huge implications on the program success

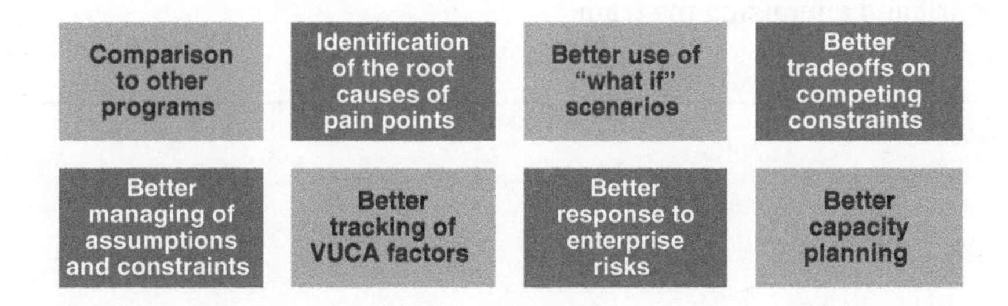

FIGURE 1.24 IOT and Program Change

1.10.4 Beyond Agility

This PMI pulse of the profession study, as summarized by figure 1.25, highlights many critical attributes that are seen affect the future capabilities of program and project managers in their handling of change.

- Continuously adapting
- Addressing the ability of the enterprise to stretch itself and be adaptive
- Finding ways to stretch and tilt and to do whatever is needed
- Having a balanced technical and business acumen
- Going beyond the agile mindset and transparency needed for the future culture
- That is collaborating on steroids
- This allows the program managers to play their most suitable roles in connecting the teams, driving change, no matter how uncertain the conditions might be

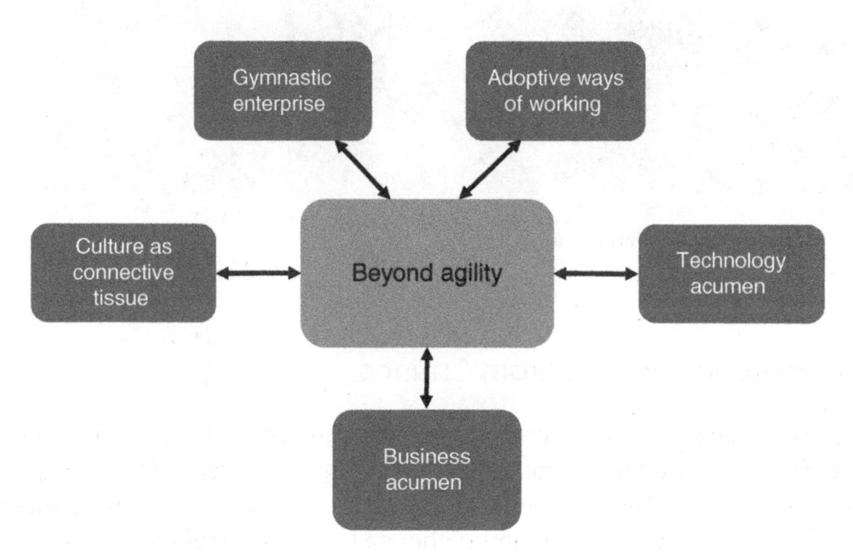

FIGURE 1.25 Going Beyond Agility
Source: Adapted from PMI 2021 Pulse of the Profession Report.

Review Questions

Parentheses ( ) are used for Multiple Choice, when one answer is correct. Brackets [ ] are used for Multiple Answer, when many answers are correct.

How does servant–leadership contribute to the success of the program teams?

( ) Participating in change management efforts.
( ) Demonstrating respect for everyone.
( ) Inspiring team spirit and energizing the team.
( ) All of the above

How does the IOT contribute to program change success?
Choose all that apply.

[ ] Fixing assumptions and constraints.
[ ] Identification of the root causes of pain points.
[ ] Better response to enterprise issues.
[ ] Better tracking of VUCA factors.

What was highlighted in PMIs Pulse of the Profession study?

( ) Structured enterprise
( ) Stable way of working
( ) Program management acumen
( ) Culture as the connective tissue

CHAPTER 2

Creating Focus

PROGRAM MANAGERS SUCCESS

These leaders are life-long learners, and they have the foresight to use their programs and projects as the learning backdrop to continue their excellence journey. The know that experimentation does not hurt and that they could learn a great deal about customers, users, and other stakeholders that would enable stronger engaging and ultimately achieves the joint focus needed.

Creating focus is a critical muscle for the program managers, the program teams, and for the extended organization. This is what enables organizations to operate strategically and create a well-connected entity of success partners. In this chapter, creating focus is tackled in multiple ways, yet most center around the engaging and aligning of the right stakeholders on the direction to take and then the needed steps to get there.

Key Learnings

1. Understand the role of the sponsor as the program coach and captain
2. Learn how to develop the important partnership with the program sponsor and the related political acumen skills
3. Explore the principles of effective communication and learn from a complex program example how to connect across tiers of stakeholders
4. Understand the importance of using value in program prioritization
5. Develop a strong awareness for the importance of crating time to intentionally think as a program manager

2.1 THE PROGRAM SPONSOR

Programs and projects benefit from having the role of a sponsor. Not just in the literal sense of having the person or the organization that is contributing the funds or the resources to execute against the program plans, but also mainly the person or the entity that stands behind the program/project mission and is there to handle any obstacles on the way, as needed.

Program Management: Going Beyond Project Management to Enable Value-Driven Change, First Edition. Al Zeitoun.

Program sponsors are not just born with it. There is a learning curve for that role. There are expectations to be understood, partnerships to be developed with the initiatives leaders, and tough decision to be made.

Key Learnings

- This has huge impact on your success as a leader as highlighted by various research and experiences
- The go to person we call the sponsor and the one thing that you need to have more than anything else for the success of your program, *having the right program sponsor*
- You might encounter the opposite to the sponsorship like in the case of meddling
- We talk here about the persons who are ready to roll up their sleeves to help you, yet are totally balanced in their involvement

This part of the chapter will further address many of the critical learnings necessary for the sponsor role to be most effective:

The role of the sponsor in adapting and progressing

- Who is that sponsor?
- Why is that role critical to your success?

Finding the right position for the involvement needle (the captain position)

- Moving toward the captain position

The coaching language and its value for influencing program success

- It is a highly needed future quality and attribute in what you do
- Critical for the success of your team
- Coaching language matters

The GROW Model applications in creating the right program culture

- Bring the application of this model to the creation of an excellent program culture[1]

2.1.1 Role of the Program Sponsor

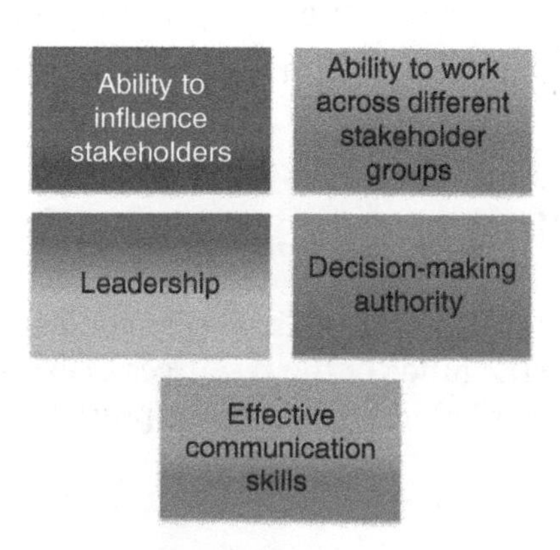

FIGURE 2.1 Attributes of the Sponsor Role

[1]Program Sponsor Paper: https://pmworldjournal.com/article/the-connected-future-business-culture.

- Figure 2.1 highlights multiple attributes of the program sponsor, centering around the true leading by example phenomena
- As a servant leader, this role helps in influencing other stakeholders, works the politics, agendas, creates alignment, and breaks down silos
- Ability to enhance the governance with clear authorization
- Ensure we have a program charter in place including knowing exactly what needs to be delegated to the program manager
- Crystal clear communication quality as it reflects directly on the team
- Adopts to the needs of supporting the success of the program manager and the program team
- Program and project managers could play that role
- We are not talking only about the person providing the funds, yet more the one who truly understands the business need
- This is the person who links to business value while supporting your and the team's success
- Always ready to move the program forward and removing obstacle out of the way

> **TIP**
>
> Program sponsors are servant leaders who embody the mission of the program. As great communicators, influence, break down silos, and create focus

2.1.2 The Captain Position[2]

Program sponsors strive to find the right degree of support for the program and the program managers. This is not an easy task, and they could err on either side of an extreme position. Whether they end up being too invoiced, almost meddling, in every detail, or the other end of taking too much of a hands-off position. Finding the right captain position is an art and science and does require a high degree of experimentation on the part of the sponsor.

Great program sponsors inspire their program teams and create a transparent, innovative, and adaptive program culture. They operate like a captain of a sports team. Seeing the big picture, connecting the dots, driving, letting go, and always present for their teams. Their success hinges on having real and clear expectations of their role. Program managers need to take an active role in explaining to the sponsors these key expectations.

Ultimately, program sponsors should be active on their programs, but not too active. They attend status meeting, review briefings, and provide direct inputs. Successful programs would typically have sponsors who don't hesitate in making course correction and deciding on major issues, e.g. being in half time in a sports event. Just like great captains, they delegate most decisions and inspire their program teams with the balanced servant leadership style.

2.1.3 The Coaching Mindset and Program Success

- This mindset helps in building that healthy environment in the relationship with the sponsor and in cascading that across the team
- Allows us to create the go to openness for the team from the sponsor onward across program roles

[2]Adopted from the Executive Sponsor Research Report. The Standish Group, 2013.

There Are Three Building Blocks for This Coaching Mindset

Behaviors consistently enable balance

- Enables the balance between the consistency and the adaptability attributes to ensure leaders continue to drive impactful program change

Future orientation in coaching

- A must-have puzzle piece for the focus on value
- Supports excelling in the role as a program manager

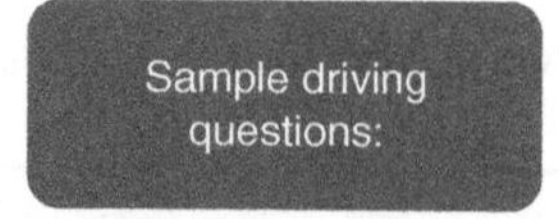

- Drive focus to value
- Raise the mirrors in front of our face

Critical Questions for Program Success

- How to evolve the role of the program sponsor?
- How to use kickoffs to address critical program questions and better organize? How it helps us in finding the key moments in the program?
- How to create urgency and build in the time to think and tailor your client approach better? Example brings back the consistent focus on outcomes?

2.1.3.1 GROW Model and Program Culture

- Highlighted in Figure 2.2, the GROW model is highly utilized tool in coaching
- Makes sense in communicating properly
- **Goal:** ensure clarity of purpose
- **Realty:** Ensure understanding of program's assumptions/constraints – true conditions of the program
- **Options:** Ensure inclusion of diverse program team members' views – multiple program options ahead of us
- **Wrap-up:** Ensure creation of call-to-action program culture – intentionality: we meet, discuss, and interact with a clear path forward

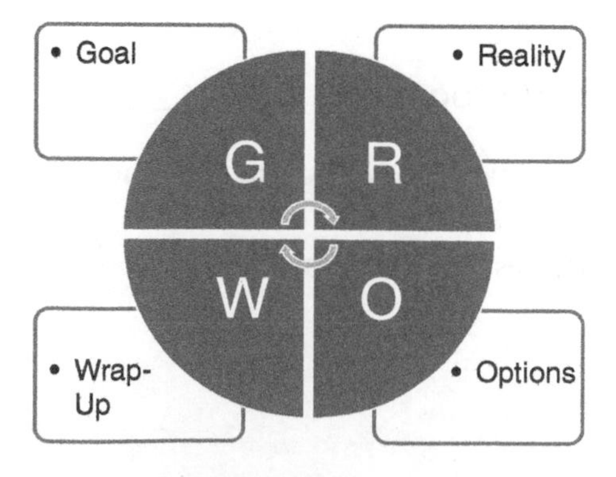

FIGURE 2.2 The GROW Model[3]

[3]Adapted from the GROW Model https://www.performanceconsultants.com/grow-model.

Review Questions

Parentheses ( ) are used for Multiple Choice, when one answer is correct. Brackets [ ] are used for Multiple Answer, when many answers are correct.

What is an attribute of a successful program sponsor?

( ) Demands reports on a regular basis
( ) Micromanages to ensure program compliance to regulations
( ) Ability to work across groups of stakeholders
( ) Available to program manager at program gates

Describe a captain-like sponsor scale.
Choose all that apply.

[ ] Rare involvement in decision-making
[ ] Makes course correction and decides on major issues
[ ] High, active involvement in the program
[ ] Delegates most decisions

What does the "W" in the GROW model stand for?

( ) Willingness
( ) Working fluidity
( ) Waterfall approach
( ) Wrap-up

2.2 A CRITICAL PARTNERSHIP

One of the well-kept secrets of programs success is this critical partnership between the program sponsor and the program manager. It is important to build on what it takes to be a good captain as a sponsor. It is also essential to deep dive into the relation with the sponsor, including key questions to use, and ensuring this fundamental relation to your success is invested in.

There is a critical value for the program sponsors role in managing key stakeholders' relations. There are many organizations, whether the government entities, or other businesses, where the executives prefer to be handling key priorities and conflicts with equally senior-level leaders. Whether right or wrong, if this is prevalent, the program manger would need a proper coverage from the program sponsors to make sure that they can focus on driving their programs' outcomes achievement, while the sponsors are taking obstacles out of the way. Continual investment in this partnership and strengthening the open dialogue around where the program manager needs the most help with stakeholders would give the program sponsor the opportunity to create the most valuable impact on the strategic success of the program.

TIP

Focus on building the proper rapport and strong sponsor partnership are key skills to develop.

Key Learnings

- The worst scenario – the program sponsor and managers don't work as partners "the dangerous program fallout"
- Putting together the key questions to ask the program sponsor – reviewing the importance and the quality of the questions we use
- Developing the muscles for timely critical conversations – not only do we expect leadership by example but also drive the excellence culture and having those key dialogues take place
- Understanding the art of cascading delegation – building on previous review of authority and governance topics
- Exploring the use of political acumen in practice – key fundamental quality to invest is centered around getting yourself involved in critical growth experiences

2.2.1 Importance of an Effective Sponsor

- Consistent data points across studies and research
- Important reminders of the value, and why we invest in this critical sponsor relation

2.2.2 The Dangerous Program Fallout

Investing in this partnership is clearly important, as seen in Figure 2.3. Continually nurturing this relationship could help avoid improper consequences.

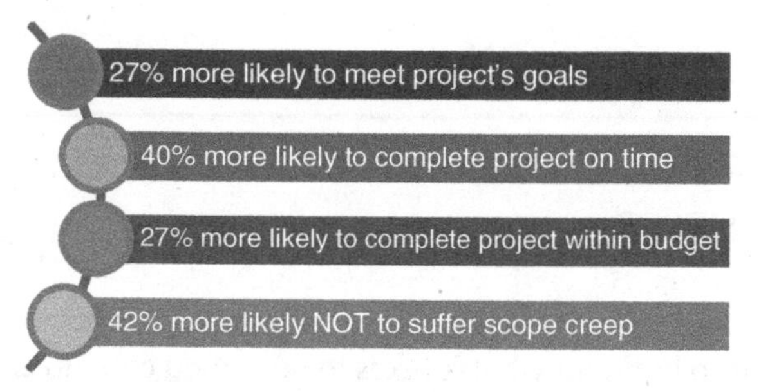

FIGURE 2.3 Impact of Effective Sponsorship[4]

Improper Partnership Consequences

- No clear purpose for a program or a well-developed program charter
- No coverage for intimate knowledge – the connections that the sponsor has
- Missing clarity of "end game" – outcome of the program becomes the gap
- Unclear path for program escalations – program manager would not know where to go
- Gaps in authorization of program manager creating roadblocks – what decisions can the program manager make and why?
- No streamlined involvement
- No one keeping their "eye on the ball" – program manager would be missing the real connection to the business
- Lack of timely intervention when necessary – to accomplish the sensitive and proactive engagement in what matters for the program's success

[4] "Executive Sponsor Engagement: Top Driver of Project & Program Success." *Executive Sponsor Engagement | PMI Pulse of Profession*, www.pmi.org/learning/thought-leadership/pulse/top-driver-project-program-success.

2.2.3 Critical Questions

> **TIP**
> Program and project managers know the value of questions. They continually develop their capability of asking the right ones timely.

The critical questions to address with the program managers relate to their roles expectations:

1. Participating in key meetings – defining what that is, controlling meddling
2. Supporting program manager:
   - business topics
   - key conflict resolutions
3. Creating relations with other sponsors:
   - have cross-programs collaboration – what is important to the sponsor?
4. Develop good understanding:
   - talk the "same language" – same wave length, no gaps in understanding the direction and key risks
5. Take time to talk/invest in developing people – most critical as it helps in investing in continual growth, such as the use of the coaching skills to help the entire program team
6. Sell program across business

Albert Einstein said, "I have no special talents. I am only passionately curious." This is such an important quality in the role of the program sponsors. Timing of the questions and how this curiosity is exemplified are also everything in the sponsor role:

- Critical muscle for agile way of working – openness in the environment
- Sets culture of transparency necessary for program success
- Allows for powerful questions: these questions need to flow and be followed by the sponsor and the program manager, enhances the quality of program outcomes:
  - Is there another way of looking at this?
  - How best could I help you?
  - What would happen if . . .?
  - What could we do to make this situation better?

2.2.4 Art of Cascading Delegation

In Figure 2.4, some critical leadership dimensions are highlighted for the role of the sponsor and the continual development of that role, together with the partnership with the program manager are highlighted. Given the importance of programs, as strategic business opportunities, some fundamental nuggets are demonstrated in the figure, such as trust, the importance of realizing the reason where the obstacle to delegating is coming from, and the willingness to develop the next tier of sponsors, are all integrated pieces of the puzzle of the successful delegation capability:

- Built on a foundation of trust
- Need to understand what is preventing the achievement of trust
- This is also why tools that allow us to see clearly who does what are great

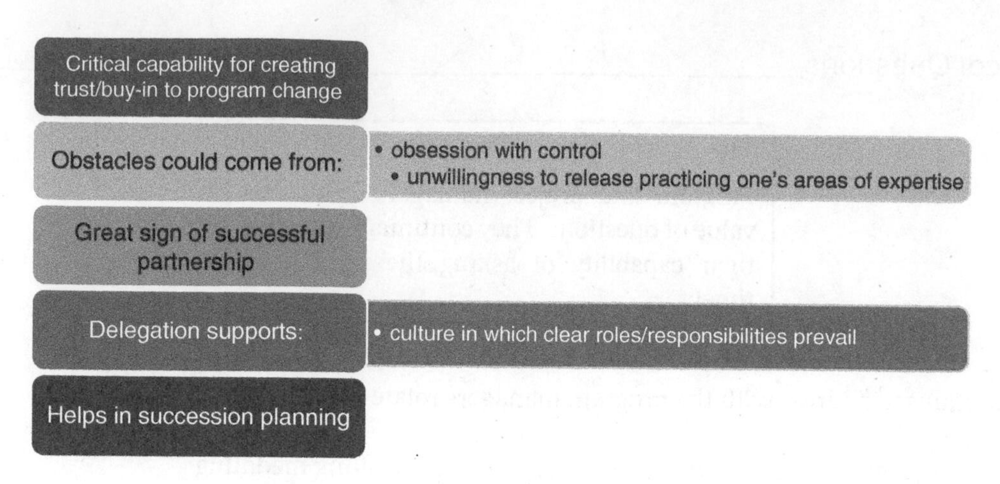

FIGURE 2.4 Building Effective Sponsorship

2.2.5 Program Manager's Political Acumen

Reflecting on politics is usually not an easy task. Politics have different meanings for different stakeholders. Most professionals are uneasy to speak about politics, let alone living with politics. It is crucial to realize that politics is part of the project/program environment, and that the leaders of these initiatives should be well equipped, not only to deal with politics but also to empower their teams with the proper channeling of politics.

Figure 2.5 shows the three key political acumen pillars: diplomacy, networking, and coupled with wisdom and support. Program managers who are able to practice these well have consistently been able to achieve program delivery excellence and protected their teams from unnecessary distractions, while creating a strong focused environment for them.

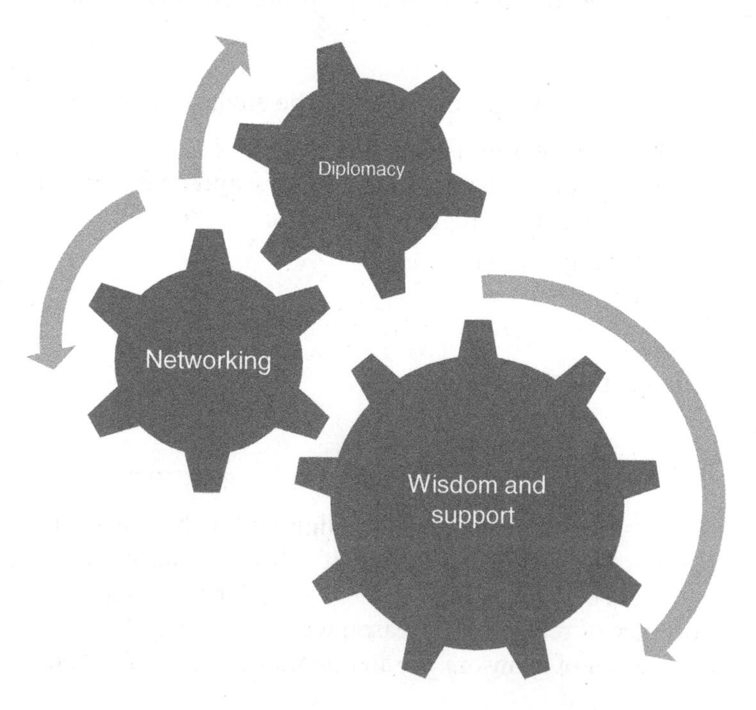

FIGURE 2.5 Political Acumen Pillars

Questions to Ask Yourself:

- Are politics in programs good or bad? Why?
  - Neither, they are key for building strong alliances focus
- How do you get to "be political" on your programs?
  - Becomes an ongoing practice of finding and repeating working patterns time and again as in the case of creating excellence

> **TIP**
> Understanding and working positively with politics is such a valuable investment of our time.

Review Questions

Parentheses ( ) are used for Multiple Choice, when one answer is correct. Brackets [ ] are used for Multiple Answer, when many answers are correct.

What is a sign of a weak partnership with the program sponsor?

( ) Attendance of key meetings
( ) Intimate knowledge of the business
( ) Unclear path for program escalations
( ) Keeping the "eye on the ball"

How does proper delegation lead to program success?
Choose all that apply

[ ] Program teams have more fun
[ ] Makes program roles and responsibilities clearer
[ ] High level of involvement in technical details
[ ] Supports succession planning

Are program politics good or bad?

( ) Good
( ) Bad
( ) Neither

2.3 THE STAKEHOLDER LINK

Establishing a strong positive link to program stakeholders creates the focus on what matters most to these stakeholders. A real favorite capability as it could take years to understand the importance of stakeholders in programs. It is such a common-sense topic for project and program managers, yet we miss taking the time for it. The real challenge is the gap between just knowing who the stakeholders are and dedicating the time that is required to invest in building this critical link. This link becomes more critical when the complexity of the program and the diversity of the stakeholders increase.

Key Learnings

- An example of using the World Health Organization (WHO) Principles for effective communications to create global impact
- A UAE Complex Program example for illustrating Stakeholders' Tiers – investing in the various tiers and the alignment building
- Applying the roles of Program Stakeholders Register to build strong program engagement – as an example of multiple tools that you can use to ensure a structure and a methodical way are being used
- Tactics in handling resistance – programs are about change and thus this brings us back to the running and changing the business dialogue and finding the tactics that maintain focus on accomplishing change

2.3.1 WHO Communications Principles

As highlighted in Figure 2.6 that summarizes the WHO principles for effective communications, these principles apply well to the world of programs and to the practices of program managers. Every one of these five dimensions would make a difference in the quality of program communications and the associated outcomes program managers could achieve by practicing these:

- This is not the day and age to be hiding information
- What can I do with the data?
- Is the data vetted, clarity, biases out of the way?
- Is the data helpful to the key strategic direction and objectives of the program?
- Communication intentionality is making sure that it hits the nail on the head and that is also why being timely is extremely important

FIGURE 2.6 Communications Principles[5]

2.3.2 Sample UAE Program and Stakeholders Tiers

Reflecting on a complex United Arab Emirates (UAE) program, in this case in the nuclear industry, one could learn how a large mix of stakeholders, e.g. the organization that is building the plants, the one with the responsibility for operating the plants, the industry regulators, the stakeholders behind bringing in the technology, and supporting the innovations, supply chain, etc.

One would start seeing the wide diversity of stakeholders. Naturally, these stakeholders have different needs, expectations, and maturity. For such a high visibility program, massive continual growth of stakeholders' focus is needed, given the commitments being placed on such a complex program with a country critically impactful program like this.

[5]Adopted from the World Health Organization communication principles (https://www.who.int/about/communications/principles).

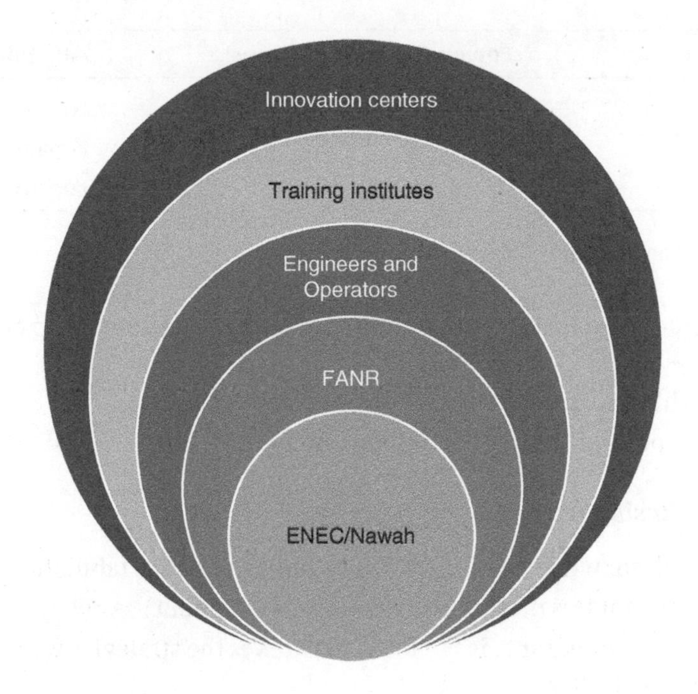

FIGURE 2.7 Tiers of Stakeholders Example

Using a cadence of continual reviews of stakeholders' engaging strategies and having the program manager working with a small core team, is highly valuable in ensuring that there is an effective workable plan to get the stakeholders best aligned around the focus of the program. Thinking through the stakeholder placement in tiers could be beneficial in gauging the level of effort and time dedication needed for these different tiers.

As emphasized by Figure 2.7, this all confirms the need to have a structured approach to work with this diversity of stakeholders. "A Coalition of Stakeholders."

Coalition of Stakeholders

1. Nuclear plant owner (ENEC) and operator (Nawah)
2. Independent regulatory body (FANR)
3. Nuclear technology center to spur innovation
4. Institutes at vocational/university levels offering nuclear engineering/other related training courses
5. Hundreds of Emiratis nuclear engineers/specialists:
   - A total of 30 reactor operators certified by FANR[6]

2.3.3 The Stakeholder Register and Setting Up Engagement

Using some structure for understanding who the stakeholders are, where they stand in relation to the program, and designing the strategies to work with them are all foundational elements for the success of engaging. Figure 2.8 shows a classical and easy way to list the stakeholders in the first column and then follow that with three dimensions that help in understanding their position toward the program. These three dimensions are *power* (which reflect the degree of influence the stakeholders could have on the program, *interest* (how much do they care and/or are concerned about the program and its outcomes), and finally even *attitude* (as it reflects their stance regarding how likely we truly get their support).

[6]https://agsiw.org/nuclear-power-in-the-middle-east-the-politics-of-stakeholder-coalitions

Stakeholder	Power	Interest	Attitude
Program Sponsor	H	H	Positive
CIO	H	L	Negative
Users	L	H	Positive

FIGURE 2.8 Sample Stakeholders Register

- Simple tabulation structure to list the power and interest and you could also combine the attitude in your assessment
- Negative attitude qualifies to bring them in the resistance category
- This is also where our political acumen development helps

Tactics for Handling Resistance

- The analysis in the grid, shown in Figure 2.9, builds on the previous tabulation that the register highlights
- We could place the stakeholders on this grid (like dots throughout)
- What we see written in the quadrants is the focus that drives the strategies we will be taking with the stakeholders to engage them properly
- Few engagements could be focused on the highlighted quadrant (top left), given the possible resistance coming from stakeholders in that quadrant and the danger of them being not engaged, putting obstacles in the way, etc.
- The program manager should engage/influence that group or protect team
- Top right is a natural continual focus group of stakeholders
- Bottom right quadrant has good interested stakeholders to build on
- Bottom left is a chance to minimize my time investment there
- All this strategic analysis helps me be more intentional in my communications building on the WHO principles earlier in the lesson
- This is a dynamic exercise so I have to revisit that effort throughout the program lifecycle
- Always pay attention to the changing conditions surrounding these stakeholders

> **TIP**
>
> Understanding your program stakeholders and engaging with them is a strategic competency that should be practiced and developed.

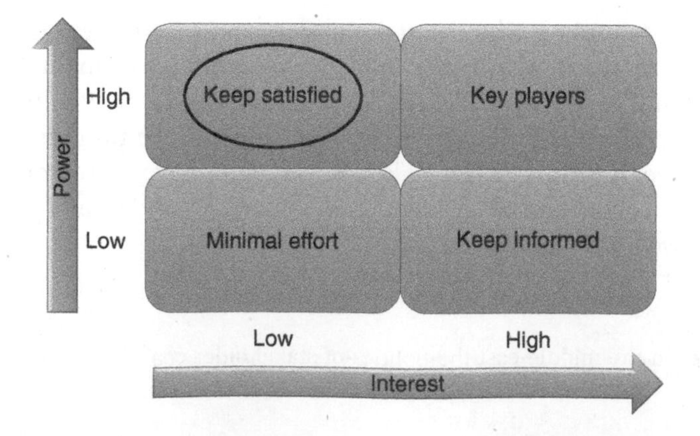

FIGURE 2.9 Stakeholders Engagement Matrix

Review Questions

Parentheses ( ) are used for Multiple Choice, when one answer is correct. Brackets [ ] are used for Multiple Answer, when many answers are correct.

Which of the following is emphasized by the WHO principles for effective communications?

( ) Any source of information is acceptable
( ) Mainly focus on conceptual topics
( ) Have a preference for more details
( ) Relevancy to the given stakeholders

What direct benefits could one expect from a well-developed stakeholders register? Choose all that apply.

[ ] Ability to prioritize the efforts needed to engage the stakeholders
[ ] Decide on the overall themes of engagement strategies
[ ] Become popular with program stakeholders
[ ] Manage resistance to program direction

What engagement tactic is typically used for stakeholders that have high influence over the program, yet are not interested enough?

( ) Minimal effort
( ) Keep informed
( ) Key players
( ) Keep satisfied

2.4 THE PROGRAM CHARTER AND CLEAR PRIORITIZATION

Having reviewed the program charter as an instrument for creating focus, another key valuable use of the charter is enhancing prioritization clarity. Regardless of the delivery approach of the program, and whether it is more on the waterfall (phases) methodology side or the much more adaptable agile delivery side, there is a very evident need for crystal clear prioritization. Charters are great at that and at documenting what the program sponsors or product owners reach in terms of that important program decision.

The program charter is crucial in the arsenal of tools that a program manager. Given that the two words that come to the surface most in the role of the program manager are communication or prioritization. Most program/project tools, even software for managing programs, are all about communicating something pertaining to the progress, engagement, and/or what takes priority over something else.

The charter provides a strong link back to the program governance role. Prioritization has an art and a science to it. This also provides clarity of where you are spending your time as a program manager. Charters help us zoom on what will count most.

Key Learnings

- Expanding the discussion of the Program Charter's Value
- How are program priorities governed and why?
- Exploring the value of core and extended program teams in maintaining program focus
- Defining the use of the charter in handling critical escalations – clarity of authorization to ensure there is a clear path for the timely program work escalations

2.4.1 The Charter's Integrating Value

As per the PMI standard or program management, program charter builds on having a proper program business case. The charter articulates nicely the reasons behind the work of the program and the anticipated benefits. Having a well-developed charter then leads to being able to envision the program roadmap that connects the dots across the program components.

The program charter establishes a framework by which program components are managed and monitored:

- The center of the universe in governance
- The charter captures the key elements
- Strategic in nature
- Driven by the program sponsor
- Clarifies what success looks like
- Environmental factors and the risk strategy importance will continue to take more value in the future of the organization
- The charter contributes directly to what goes on the program roadmap
- All of this contributes to ultimately having a solid and comprehensive program plan
- This is key to integrating our focus and managing our key priorities

> **TIP**
>
> Have the charter help you and link you back to the sponsor to ensure clarity of governance across the team and extended team.

2.4.2 Priorities Governance

- Governance, highlighted in Figure 2.10, aligns with benefits realization – charter facilitates that
- Governance is affected by: value based
  - culture
  - maturity
  - risk tolerance – an important topic that continuously gets revisited by the program team
  - strategic objectives – e.g. alignment to regulations and other constraints too

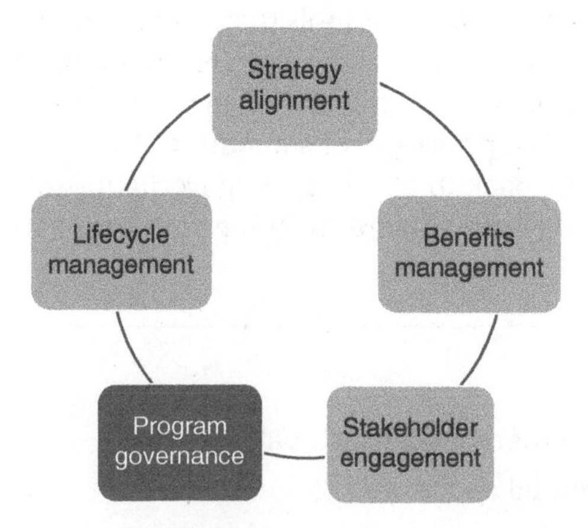

FIGURE 2.10 Program Management Performance Domains
Source: Adapted from *The Standard for Program Management* – Fourth Edition, p. 67.

- Governance adapts throughout program lifecycle – not one style of governance across the phases especially due to the changing nature of stakeholders across the lifecycle, adapting is critical

> **TIP**
>
> Creating a value-based program governance that has the right risk tolerance, enhances prioritization, and benefits realization.

2.4.2.1 Program Teams Structure and Focus

As highlighted in Figure 2.11, another key element in clear governance is the proper set up of the program team's structure and its focus. As in the UAE example earlier in this chapter, having a core team and then others working surrounding that team directly contributes to alignment and to better prioritization opportunities.

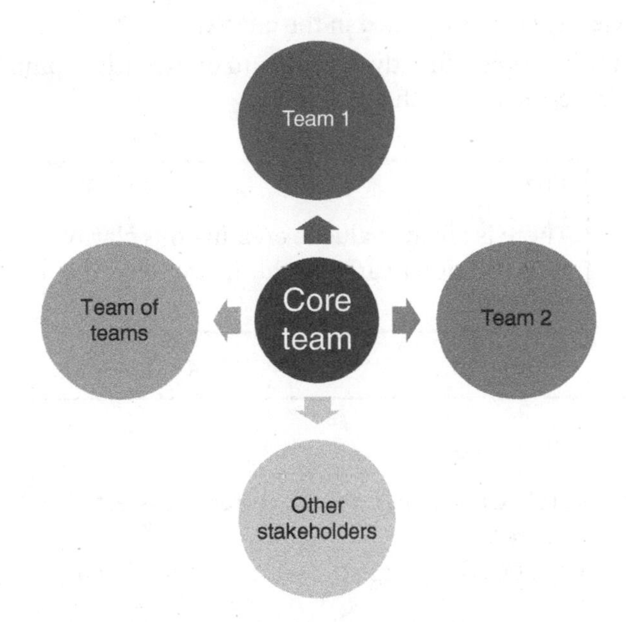

FIGURE 2.11 Core Team and Team of Teams

Representation of each of the teams is part of the team of teams as an additional opportunity to streamline the governance and the decision-making. Flexibility of team formation is important to better map and to address the program delivery approach. In addition, the degree of complexity of program contributes to selecting right structure. The program manager should consider geography/global/virtual elements in designing right team model for aligning focus.

Making sure that all above factors are taken into the mix to streamline the most suitable governance and decision-making process.

2.4.3 The Charter and Critical Escalations

Programs encounter a need for escalation. The charter is a handy instrument to support the thresholds that guide the process and highlight the different stakeholders and entities involved. Figure 2.12 shows these various potential escalation lines together with the associated best practices for these stakeholders' involvement.

- Core benefit in the use of the program charter
- Revisit the role of the charter in driving key and clear escalations

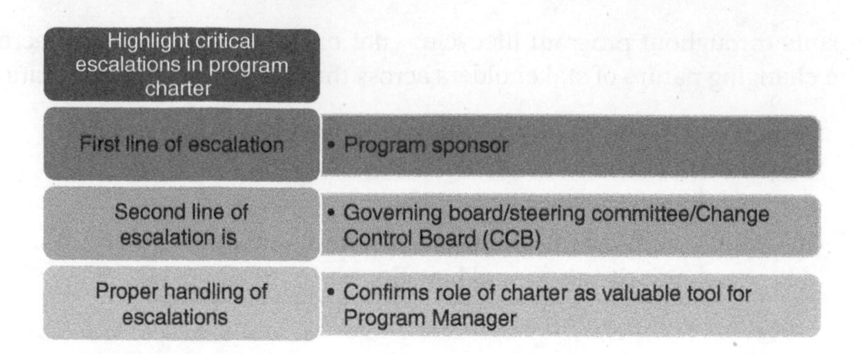

FIGURE 2.12 Program's Critical Escalations

- The escalation lines are from the program manager to the sponsor and what they bring in business clarity, then as needed to the CCB for the changes that are above the authority of the sponsor
- All of which should be clearly communicated in the charter
- Program manager might bring topics directly to the board or executive committee depending on the nature of the organization and the criticality of the program

> **TIP**
>
> There is always value to creating this clarity of escalations early in the lifecycle of the program.

Review Questions

Parentheses ( ) are used for Multiple Choice, when one answer is correct. Brackets [ ] are used for Multiple Answer, when many answers are correct.

What elements of governance does the program charter bi-directionally integrate?

( ) PMP and strategic plan
( ) Risk strategy and environmental assessment
( ) Program business case and roadmap
( ) Program roadmap and risk strategy

What contributes to the design of the core and extended program teams' structure? Choose all that apply.

[ ] Program delivery approach
[ ] Program complexity
[ ] View of the program sponsor
[ ] Virtual spread of the teams

Where would a program charter typically indicate the program would go first to handle critical issues?

( ) The resource managers
( ) Other program managers
( ) A change control board
( ) The program sponsor

2.5 THINKING AGAIN FOR A CHANGE

What a common-sense concept and what a title we should have in front of our eyes as leaders in this ever-increasing complex and noisy way of working across the globe. It is almost sad that we even have to talk about it. Even though we are in a decade that deals with mega events, wars, dealing with crises and pandemics, you wonder how is this missing.

Creating a knowledge management culture is so important and thus is the lessons learned, retrospectives, and other related ceremonies. When working with executives sometimes they can't even answer the question of where they are spending their time. This makes the protecting of thinking time and being present to where program stakeholders focus, a high priority topic.

Key Learnings

- Reflecting back on the Conductor role's value for enabling thinking across the team
- Understanding the practices behind thinking fast and slow – an opportunity to see how that can help you think differently
- How to overcome the illusion of understanding to minimize program decisions biases? – The danger associated with making the wrong assumptions
- Identifying practical tips for creating time to think again as Program Managers – we need to demonstrate the right examples to the program team and many other stakeholders

2.5.1 The Conductor's Role in Enabling Thinking

Figure 2.13 brings us back to the analogy of the conductor earlier. This highlights the importance of the harmonization the conductor creates. Just like in the orchestra, conductors remind us of how to be thoughtful and more aware. In a program environment, the conductor facilitates safety in the program environment and the ability to build the distance to think again, while using data insights.

When program manager operates as conductor, the strategic question is

Being thoughtful/aware of how we think directly improves

- Program decision-making
- Effectiveness of solving complex problems under increasing pressures

FIGURE 2.13 Conductor Driving Thinking
Source: geralt / 25597 images.

2.5.2 Thinking Fast and Slow

- Two systems, are tabulated in Figure 2.14, the fast and slow
- System 2 allows us to slow down and breathe
- Separating the automatic reactions from the proactive work we need to do more of
- System 2 allows us to step back, e.g. not quickly hit send on that e-mail
- Key to take the time to think especially in doing work that had implications on human life or other critical regulations
- Need to foster this understanding in how we think about our work and in how our team thinks too

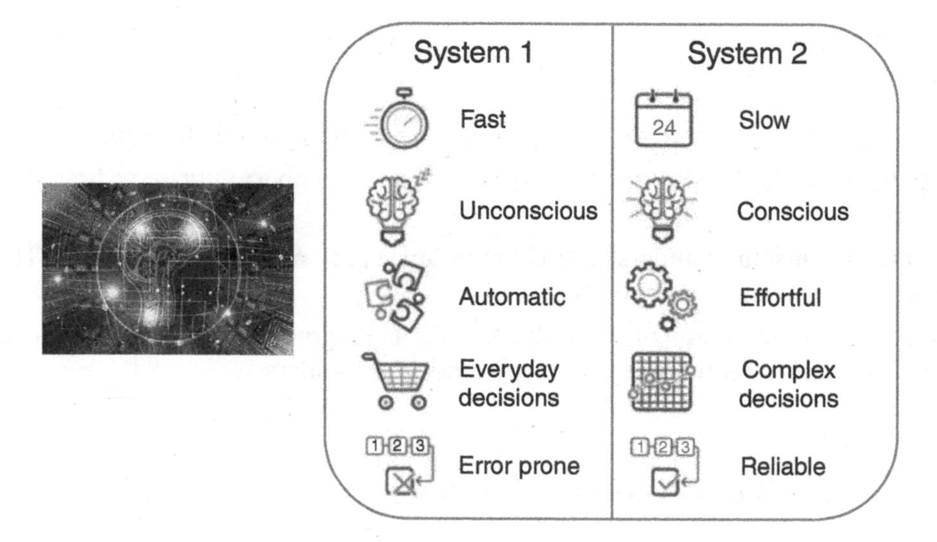

FIGURE 2.14 Adoption from Thinking Fast – Slow[7]
Source: geralt / Pixabay.

2.5.3 Program Decision Biases

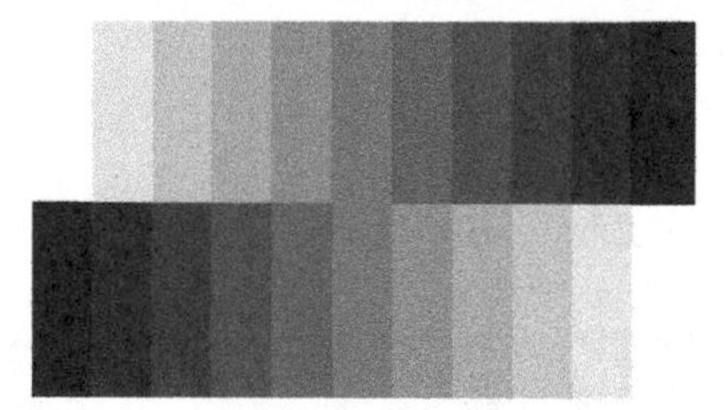

As the shades scale indicates, program decisions are not an exact science. No matter how much data analytics we have to support the decision-making process, there are many factors involved in decisions process. The biases could also weaken the quality of these program decisions.

Decision bias

- "either/or" (e.g. black or white)

Solving problems/handling conflicts/evaluating major risks

- Reset our mindset toward a balance position – gray

[7]Adopted from Thinking Fast-Slow, by Daniel Kahneman.

Grayscale

- ideal strategic spot
- achieve win-win decisions

Supporting Details

- We discussed this topic in the ABCD tool review before
- Program choices are not always clear as black and white
- The more complex the program is, the more you will have situations that have politics and other things fostering behind the scenes
- Need to be connected to the roots behind a certain decision
- Reset to a balanced position such as we did in the program sponsor being a good captain
- The goal is to reach an ideal strategic spot
- Remember the program competing constraints are not only the classic ones but more focused on the value and strategic purpose – this matters the most in making our decisions

2.5.4 Creating Time to Think Again

The roadmap to thinking again highlighted in Figure 2.15 highlights key reminders for the program managers, their program teams, and the extended program stakeholders:

- Important headline to have in front of our eyes
- Use and mark your calendar
- This understanding of time management and illustrating how you do that is critical in demonstrating this priority to your program team
- This strengthens our ability for making tough choices, saying no, cancelling unnecessary meetings, and staying intentional
- Allowing us the distance such as getting on the balcony, the orchestra view, the pausing, etc., all this makes us make different and better choices
- Prioritization around these tough choices creates the successful patterns time and again and links back to the continual impact of creating change excellence in organizations

> **TIP**
>
> Creating time to think again sounds easy, yet it is quite difficult to do.

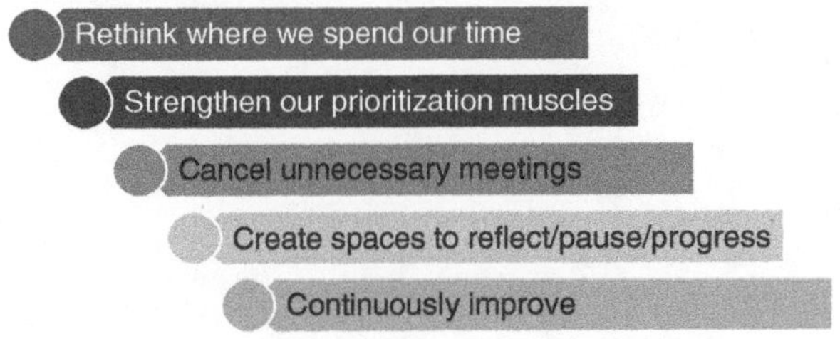

Rethink where we spend our time
Strengthen our prioritization muscles
Cancel unnecessary meetings
Create spaces to reflect/pause/progress
Continuously improve

FIGURE 2.15 Think again Roadmap

Review Questions

Parentheses ( ) are used for Multiple Choice, when one answer is correct. Brackets [ ] are used for Multiple Answer, when many answers are correct.

Which thinking system contributes to the strategic and transformational thinking of the program manager?

[ ] System 4
[ ] System 3
[ ] System 2
[ ] System 1

How does the conductor best support the program team's need to think again for a change?

( ) Apply system 1
( ) Gaining the program management professional certification
( ) Rethinking the program priorities
( ) Use daily meditation sessions

How can you best achieve a win-win stance in resolving programs conflicts?

( ) Stick to your perspective on the program
( ) Give in to allow for smoothing the conflicts
( ) Try to develop a black or white position
( ) Target the grayscale zone and practice that balance

CHAPTER 3

Driving Integration

Program managers are integrators of people, processes, systems, technology, and most of all execution priorities. As highlighted in previous chapters, this integration role of the program manager is fundamental to a program's success. The program manager connects the dots, conducts the business of the program, and aligns the troops toward the purpose and the ultimate achievement of program benefits. Driving integration requires the program manager to continually learn and practice creative ways to deliver this aspired effective integration value.

Key Learnings

1. Understand, using a case example, the ways to become a successful holistic leader
2. Get immersed into the importance of the program benefits focus and the use of enterprise risk management in supporting the benefits achievement
3. Learn from thought leaders' views on how to use empathy in making courageous decisions and strengthen your execution discipline
4. Learn the art of using program roadmaps to achieve team connectedness and simplify the focus on programs' value achievement
5. Develop your storytelling appetite with examples across radio guidance, agile ceremonies, and program canvas

3.1 THE HOLISTIC LEADER

For the program managers to become such effective integrators, there are multiple ways to tackle this integration, and becoming and practicing being a holistic leader is one of them. This allows you as program managers to understand the importance of integration, at the program level, by looking end-to-end, developing better thinking muscles, and creating the objective distance required to be holistic, such as when we addressed previously the tool of being on the balcony versus being on the dance floor.

Program Management: Going Beyond Project Management to Enable Value-Driven Change, First Edition. Al Zeitoun.

This helps shape your focus on becoming a more capable strategic leader. At the end of the day, this holistic view quality is fundamental not only to the success of the program, but also to the true value you can provide your program team by observing and learning valuable information and achieving better context understanding.

Key Learnings

- How could the leader bring the program team members to their flow? – e.g. coaching and it is a commitment that you must have as a program manager
- Use the program lifecycle to create an integrated program delivery approach – another key integration vehicle
- Use the Disney Imagineering example to demonstrate the connection between people and technology and then create unique results – what happens when we utilize the true connectedness to the customer, another reminder of the focus on outcomes
- Balance the business and strategy understanding with the integration process – program success hinges on how much we are keeping benefits at the center of our focus

3.1.1 Getting Program Team Members to Flow

A critical part of driving integration is the ability of the program managers to facilitate getting their program teams members into their flow. This success of that effort is witnessed when we see the teams' roles and responsibilities are clear, motivation is high, and work is being done in a harmonized fashion. Flow enables the best use of the different skills and views on the team and getting the forward movement focused toward programs value. There are a few things to think about regarding the importance of this point:

- Ensuring the right people are on the bus – skillsets and capabilities
- Sitting in the right places on the bus
- Flexibility in the roles and responsibilities with ownership clarity – tools such as matrices
- Developing T-shaped skills – this is so important to the future of programs success, it is investing in the depth of skills and expertise, yet also investing in the complementary skillsets that contribute to your success, allow you to shift people around, speak the same language, and create the continual focus

The PMI Standard for Program Management highlights a set of supporting program activities that help the program managers in getting their team into flow. A few of these supporting activities are similar to what the project managers are accustomed to when they prepare themselves with the knowledge areas needed for planning and driving their projects success such as scope, schedule, quality, risk, resource, procurement, and communications management.

On the other hand, program information management, program financial management, and program change management, are the unique supporting activities that match the true needs for the program and enable the program manager to use data, financial, and other supporting change management activities to drive the journey of change necessary for their programs' success.

Program managers should further dive into the details of these supporting activities and customize what they would find most suitable to use in getting their program teams in flow:

- Take time to review the details of the specific processes that cross the supporting activities and the program lifecycle phases
- These supporting activities also reflect some of the knowledge areas critical to project and program work
- As you look across these processes, you can also determine which process are most valuable for governing the program

- For example, the risk processes and the balance between threats and opportunities they bring and thus typically a rigorous set of processes that support the focus on value
- These processes help you better manage risk and other attributes necessary for successful program delivery and ultimate closure
- Reviewing these processes and looking and reading details behind each of them in the program management standards supports your choice of what helps you most in integrating the work of your program

3.1.2 Disney Imagineering and Unique Results

When we look at examples for integration, one of the most critical values is the ability to integrate with the customers, their true needs, and what makes their experience of our product and services flawless. The Disney Imagineering, highlighted by figure 3.1, is such an example. It is an example of closely integrating the view of the customers into the mix.

FIGURE 3.1 Learning from Disney Imagineering
Source: dazbrin / 1 image.

Imagineering is the team that is in charge of building the resorts, theme parks, and other critical Disney infrastructure. The work they have done to enhance the customer experience shows unlimited thinking and the view of the potential. This is similar to the case of program management when we plan first as if we have no constraints. Creativity in their approach is enabled with the visuals and that has applicability in our program work.

- Imagineering is responsible for designing and building theme parks, resorts, and entertainment venues
- Outcomes are visual stories of the blue-sky speculation
- From Audio-Animatronics to Autonomatronics – shifting from repeating certain stories or what visitors witness to a much more sensing drive way of the visitors' experience

3.1.3 Balanced Business and Strategy Integration

Figure 3.2 highlights the importance of continually linking strategy and business. Any disconnects here affects the program managers' ability to translate the expected outcomes of their programs to their program teams. These five layers of the graph are important investments of time for the program manager and will increase

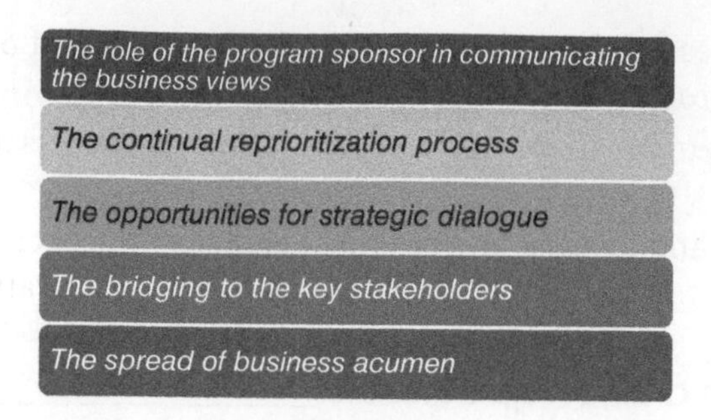

FIGURE 3.2 Balancing Strategy and Business

his/her ability to stay aligned with where the business executives heads are in terms of the strategic priorities necessary to be cascaded to the program team.

- At the end of the day, this is a nice lead into the focus on roadmaps
- The biggest link between the program and the business is the program sponsor and how close and connected he/she is
- Reminder of the two key words of prioritization and communications
- Great way to ensure strategic alignment
- Builds on taking time to think again
- The more you get exposed to the complexity that happens in integration and working across tiers of stakeholders, the more you get a chance to practice your business acumen
- This then becomes a set of widespread capabilities cross the program team and the affected businesses
- Always with an eye on aligning the program team to the strategy

TIP

Program managers should invest the time across the 5 layers of balancing strategy and business.

Review Questions

Parentheses ( ) are used for Multiple Choice, when one answer is correct. Brackets [ ] are used for Multiple Answer, when many answers are correct.

______ skills are critical for creating diverse capabilities on the program team and minimizing bottle necks.

( ) Digital.
( ) T-shaped.
( ) People.
( ) Business.

Which processes are used to integrate the elements of program delivery?
Choose all that apply

[ ] Program quality assessment.
[ ] Program risk response management.
[ ] Program risk transition.
[ ] Lessons Learned.

What is unique about the Imagineering story in the strategic role of program managers?

( ) Outcomes are visual stories of the blue-sky speculation.
( ) Insights.
( ) Digital technology.
( ) The integration of design and construction roles.

3.2 THE BENEFITS FOCUS

This part of the chapter addresses an interesting way to talk about benefits in the context of risk management. When we utilize this focus on benefits for governing and integrating better, this minimizes the unpleasant surprises that something important is missed. This focus is critically important to the success of the program teams. This also gives the program manager another review of the crystal-clear understanding of the outcomes' importance. Part of this integration is linking to the view of portfolio management and not missing enterprise level risks, being key to the program's success.

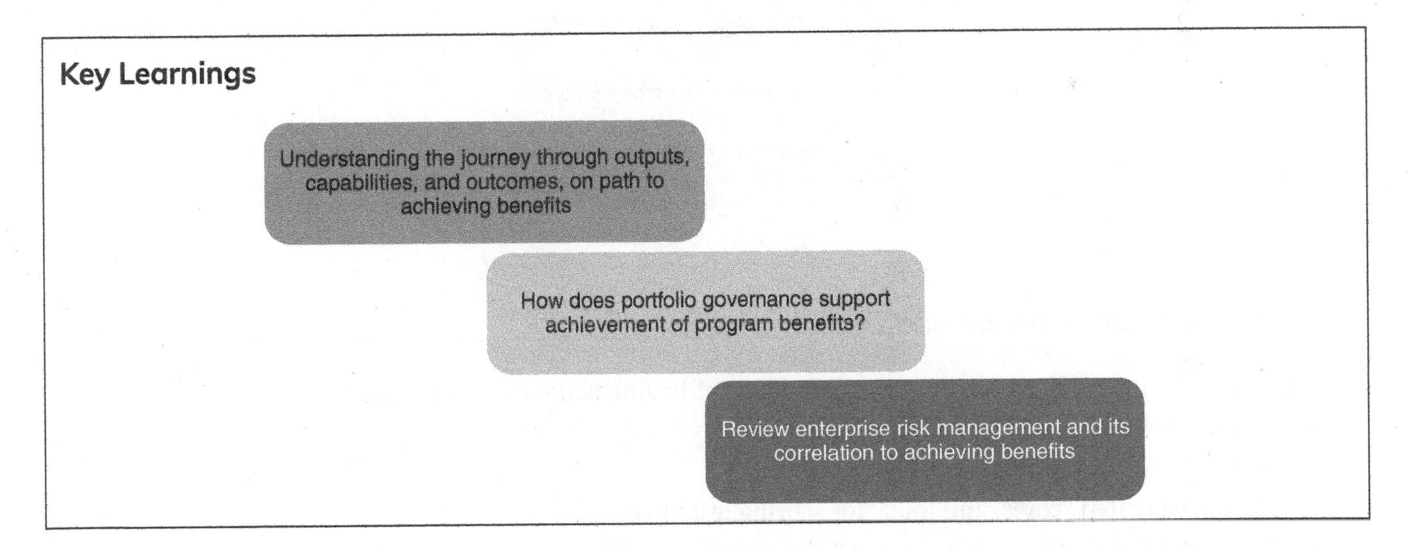

3.2.1 Outputs, Capabilities, Outcomes, and Achieving Benefits

The advantage of thinking deeply about this topic is to make sure that we are focusing on the right things. As in Figure 3.3, knowing what to measure and why, is critical to a program's success. Like in the tennis game, it is critical to keep your eye on the ball. Losing sight of that results in losing the game or in the terminology of a program, not meeting the expected outcomes. Also, the key definitional distinctions in Figure 3.4 are important differentiators that the program manager should ensure the program key stakeholders are aligned on.

- Scores, as in sports, are great, yet the final outcome matters most
- We are after the outcomes which are important in reaching the ultimate destination
- Ultimate destination is achieving value and that is what customers, users, and other key stakeholders care about
- This is a small distinction, yet could end up a critical gap in the understanding of the team and other stakeholders

FIGURE 3.3 What Should We Keep Score of?
Source: 6493990 / 764 images.

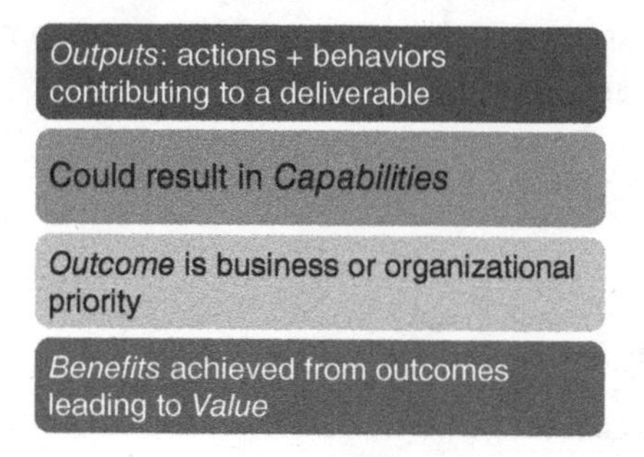

FIGURE 3.4 Key Distinctions

3.2.2 Portfolio Governance and Benefits

- Figure 3.5 shows a simple example to remind us of having benefits in our focus
- The criteria we choose will become benefits driven
- What to keep, drop, and maintain
- Criteria, weighted scores, and totals for scoring initiatives
- This is an example of one initiative, yet the concept could be used in comparing one initiative to others in terms of value

Initiative Y		**Benefit**				
	Weight	**None**	**Some**	**High**	**Score**	**Total**
Prioritized Impact Criteria						
1. Supports at least one strategic objective	20%	0	5	10	10	2
2. Realizes value within 2 months	20%	0	5	10	5	1
3. Contributes to brand	10%	0	5	10	0	0
4. Adheres to regulations	25%	0	5	10	10	2.5
5. Helps with moral issues	5%	0	5	10	5	0.25
6. Improves operational readiness	10%	0	5	10	5	0.50
7. Enhances governance excellence	10%	0	5	10	10	1
				Impact priority score		7.25

FIGURE 3.5 Weighing Benefits Sample
Adapted from "*MoP – Portfolio Management*."

3.2.3 Enterprise Risk Management

Enterprise Risk Management (ERM) is a great way to get the integration focus to the benefits mindset.

Focus on creating a common language, such as in ISO and COSO and other standards that say we should invest in risk at the enterprise level, to look at other dimensions past process such as culture.

In the UAE example, the organization created an entity focused on ERM, with the proper sponsorship like a Chief Financial Officer (CFO) in order to get the right attention and support needed for the program team.

- ERM framework could be hybrid from industry best practices
- A common language – everyone must understand terminology
- Training should be tailored to individual requirements

Figure 3.6 shows a simple illustration as an example of how to visually be in the Go position. Although this is not only a straight-line decision-making exercise, we could draw a curve that shows the most opportune moments for benefits achievement. Enterprise risk function helps in looking holistically at how the decisions are best made.

- ERM should support strategic decision-making
- Executive and senior management sponsorship and endorsement
- Change content of communications to include risk management – especially by the executive leadership leading to a level of maturity that changes the game for the success of program work

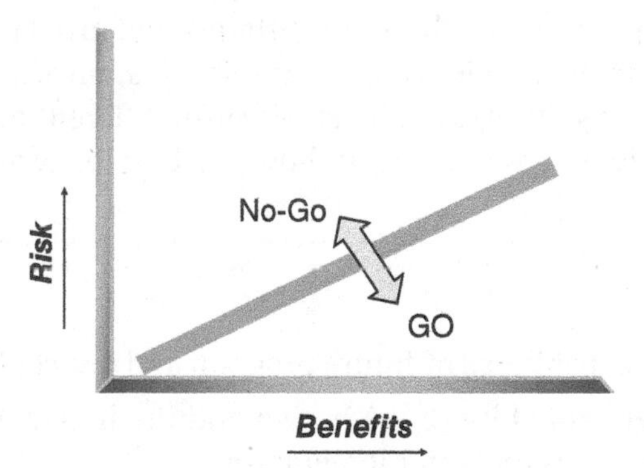

FIGURE 3.6 Weighing Benefits Sample

Review Questions

Parentheses ( ) are used for Multiple Choice, when one answer is correct. Brackets [ ] are used for Multiple Answer, when many answers are correct.

_____ are the most strategically critical in their linkages to achieving clients' value.

( ) Activities
( ) Benefits
( ) Capabilities
( ) Deliverables

(continued)

What are examples that illustrate focus on value in governing a portfolio pf program and projects? Choose all that apply

[ ] Always exceeding expectations
[ ] Speed of delivering benefits to users
[ ] Achieving more scope
[ ] Degree of linkage to strategic objectives

What is critical in creating the right culture for successful use of ERM in governance?

( ) Make the Go / No-Go process iterative
( ) Increase the focus on the speed of handling issues
( ) Ensuring the use of quantitative risk management
( ) Executive management sponsorship and endorsement

3.3 INTEGRATING WITH EMPATHY

It could be super exciting to talk about empathy and its role in creating success in connecting as a leader. Emotional Intelligence (EQ) is important to connecting the team similar to the other topics addressed earlier regarding political acumen. This gets you super ready to absorb uncertainty and dive at the deep end when you are at your best regarding topics such as vulnerability. This is not about soft skills, since as we know PMI and other organizations have been demonstrating these skills are the most powerful differentiating skills. Success is not about the latest digital twin or in AI applications, it is more about how you best relate as a leader.

Key Learnings

- Talent gap plus understanding capabilities of future program and project leaders
- Explore empathy meaning highlighted by Brené Brown and the impact in courageous program integration – Dr. Brown connects us to ways for better integration
- Develop understanding that program execution is about discipline – GM and other organizations, military, and others experienced how this is a critical component of success
- Explore vulnerability and its use to create a tightly connected program team – about creating safety and openness of dialogue and this is where the next-gen program management is heading

3.3.1 Global Program and Project Management Talent

Project Management Institute (PMI) and other organizations have been doing the work of assessing the necessary program and project management talent over the years.

- What are some of the differentiating skills?
  - System Thinking
  - Leading for Change – need to develop an appetite to be keen on including the right changes at the right time (hungry for that with an eye on priorities and other constraints)
  - Connecting Across Boundaries – culture differences and unique aspects of regions have enough distinction that requires us to cherish that diversity in our work and how this could be a unique plus for our success

3.3.2 Empathy and the Courageous Program Integration

- Brené Brown in her podcasts and videos shows the empathy definition and how this enables program managers to connect
- Critical program conversations flow better – when you have true empathy and take the time to understand, this allows meaningful conversations to happen better and often
- End-to-end connections are anchored across teams – comes from the fact that I relate to the team better, with a stronger networking, and thus encourage the openness needed, getting to the edge, making bolder moves, and reaching tougher decisions

3.3.3 Program Execution Is About Discipline

Figure 3.7 shows that in order to achieve this program execution discipline, it is ultimately about resources, energy, and time investment in what matters. How to achieve the execution discipline across these 3 buckets is broken down into the relating elements to each. In addition, shifting to more of a PMO like and other strategic moves might be considered to find what matches better what is needed for success.

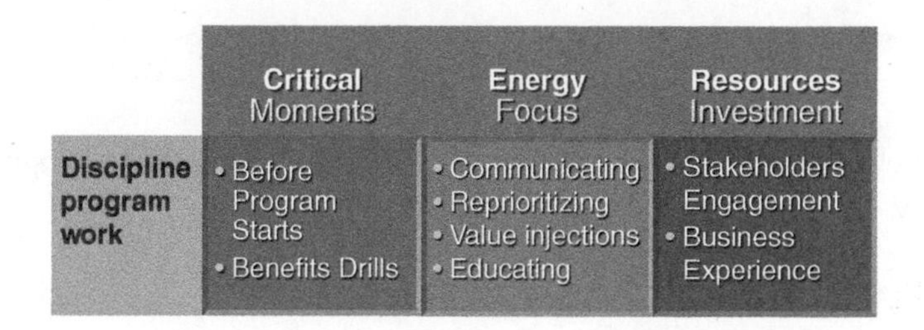

FIGURE 3.7 Execution Discipline

3.3.4 Vulnerability Matters

Another topic that is not easy for leaders and program managers to talk about is vulnerability. The degree of openness required for this in practice is high and requires a high level of self-confidence and comfort in letting go. This is however one of best strengths that you can provide your program team. This brings us back to servant leadership qualities and the importance of being humble in the approach with the team and other Subject Matter Experts (SMEs).

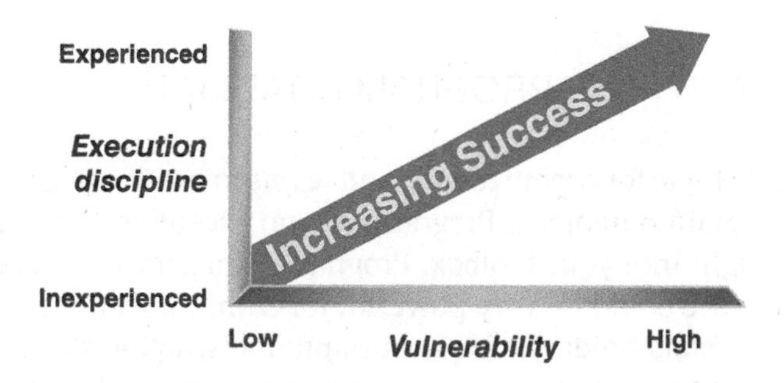

FIGURE 3.8 Vulnerability and Execution

As reflected in figure 3.8, the more experienced and vulnerable you are, the more this puts you in a good position to enable your team's success. In an ideal sense, we want to have a team that share opinions to support the knowledge management culture and the enhancement of continuous improvement and excellence focus.

- Higher impact when combined with discipline
- Vulnerability opens door to innovative inputs
- Increases potential for creating a learning culture

Review Questions

Parentheses ( ) are used for Multiple Choice, when one answer is correct. Brackets [ ] are used for Multiple Answer, when many answers are correct.

_____ skills are critical for bringing diverse capabilities on the program teams.

( ) Digital.
( ) Connecting across boundaries.
( ) Leading for change.
( ) System thinking.

Why does empathy help the program manager in successfully driving tough program team conversations? Choose all that apply

[ ] Programs are complex.
[ ] Higher chance of a true holistic program view.
[ ] Because Brené Brown identified that.
[ ] Better connection to team core issues.

Which of the following is an example of where program execution discipline is well demonstrated?

( ) Value injections.
( ) Providing multiple stakeholders reports.
( ) Detailed planning.
( ) Increasing focus on issues management.

3.4 COMMUNICATING WITH THE PROGRAM ROADMAP

Roadmaps have become a classic topic for organizations and teams to use in integrating views about steady movement toward achievement of certain outcomes. Programs are no exception to the use of roadmaps. This is also another very useful tool to go right into your toolbox. Program managers should become super creative around the idea of using roadmaps. Picture could be very powerful for communication. In this case, the program manager can use pictures to integrate stakeholders, views, work products, action steps, decision points, releases, and many additional valuable topics. There is much value you get from pictures that show flow while achieving value on the way. This contributes to making the program lifecycle well connected.

Key Learnings

- Use the Siemens case study to showcase the great acceleration that is achieved with a data rich program roadmap
- Use Simon Sinek's "Leaders Speak Last" concept to co-develop the fitting roadmaps
- Understand the key attributes of a roadmap that is designed for fit
- Weigh the pros and cons of program roadmaps – the importance of the visualization power that comes from this and the reminders of creating creative spaces

3.4.1 Role of Data in Program Roadmaps

- Richness of data that is gathered in the Siemens case[1] across the program lifecycle supports a comprehensive roadmap
- Instead of a gut feel set of decisions, we use data for strategic trends
- As reflected in figure 3.9, program changes are affected by the scientific and trending view of the data
- Trust currency is very important in our program's decisions
- The more technologically dependent we become, the more we want to invest in trust

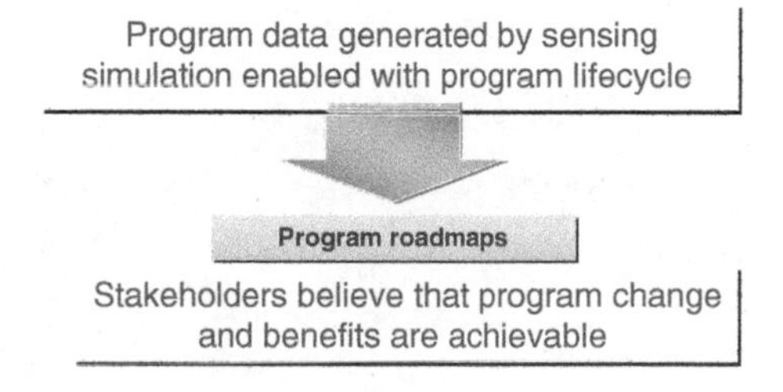

FIGURE 3.9 Data and Program Roadmaps

3.4.2 Program Leaders Speak Last

Using roadmaps to communicate is ultimately successful when the program managers are exemplifying great communications. One of these principles for successful communication is the one by Simon Sinek when he suggests leaders speak last. Simon Sinek's great stories center on starting with the "why" and in this case, it is about his view regarding leaders speaking last. This requires leaders to be comfortable with that that idea and putting value on it. Balance is critical in gauging when to jump in or not, and this capability is very important just like when we discussed the captain role for the program sponsor earlier.

The Implications of This for Program Teams:

- Importance of creating safe space
- Enhancing the trust factor
- Nourishing team learning – team learns by itself and should not be waiting on the leader to throw in a solution to the problems
- Building team diversify and connectedness – when the space is given
- Setting up value mindset – this is what we want to integrate around

[1]Link to the Whitepaper: https://resources.sw.siemens.com/es-ES/white-paper-integrated-program-lifecycle-management-consumer-product

3.4.3 Fitting Program Roadmaps

Some key features are illustrated in the sample roadmap shown in Figure 3.10:

- Now elements and future elements are illustrated
- Key strategic pillars like the 4 shown, e.g. service delivery
- Time horizons, with themes, are illustrated
- Key initiatives or benefits you want to achieve

This is almost a scientific way to do the strategic program cascade, yet you should be creative and use your own most suitable way to show movement on the program objectives.

Highlights of an Effective Program Roadmap:

- Visual
- Links to strategic targets
- Focuses on value
- Illustrates change management
- Crystallizes strategic pillars – the foundational elements of the strategy which ultimately are cascaded down to the program level and its outcomes

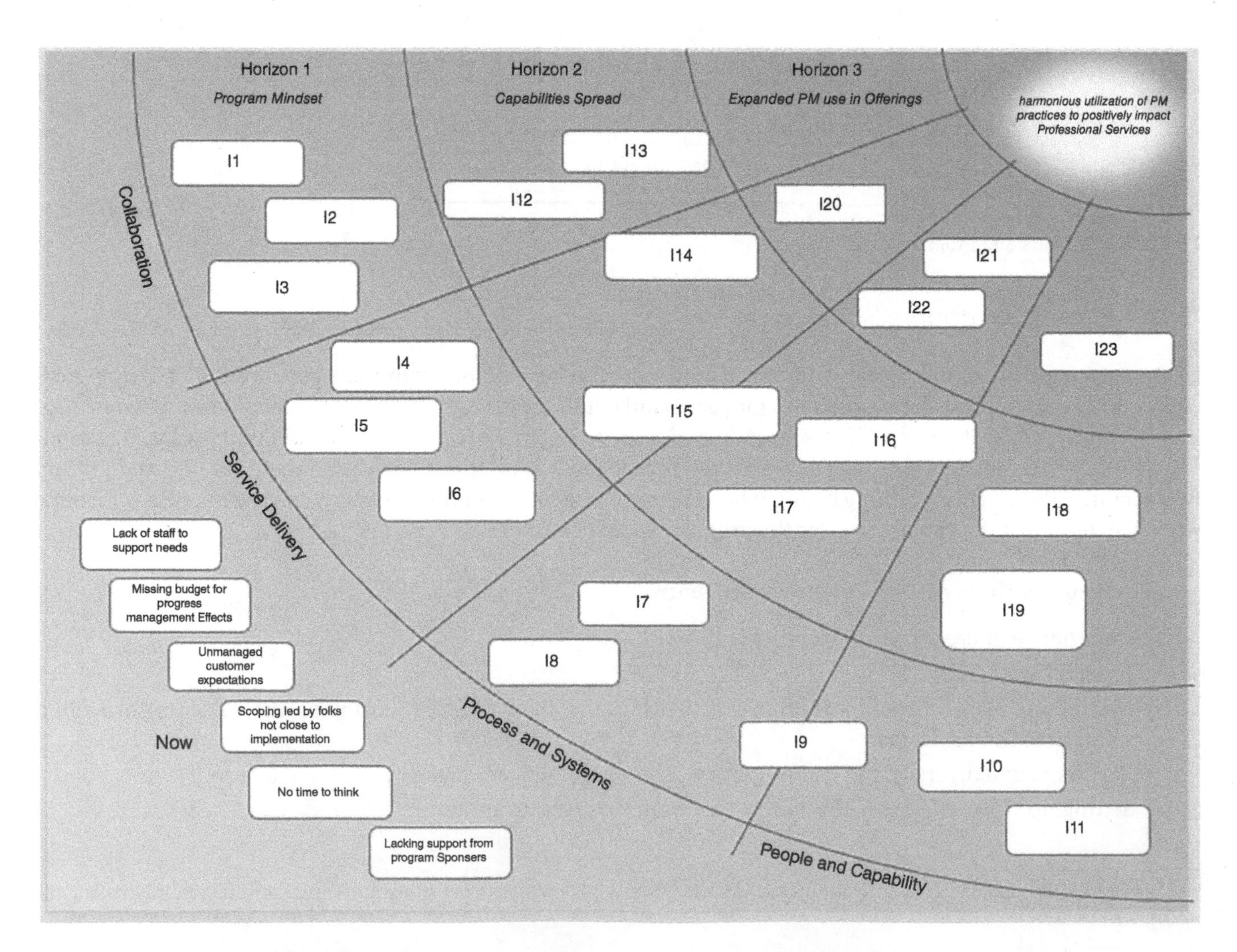

FIGURE 3.10 Sample Program Roadmap

TIP

Program managers should be creative in articulating a visual roadmap that reflects the program's progress toward achieving value.

Reflect on When to Use or Not to Use a Program Roadmap:

Use	Not to use
• When ready to shift focus to value • Sponsors commitment in place • Creating alignment is vital	• When maturity level is low • Common program language missing • Lack of outcomes understanding

Reminders

- For roadmaps, it is not an exact science
- Be careful when there is still a need to use checklists or simpler ways on the maturity path to get to using program roadmaps consistently
- Executives and other leaders need to be committed to using the program roadmaps for program's focused decision-making
- Simple topics, yet they are key in connecting the person, the process, and the roadmap's outcomes focus, which are all enablers to creating the integration necessary for the program work's success

Review Questions

Parentheses ( ) are used for Multiple Choice, when one answer is correct. Brackets [ ] are used for Multiple Answer, when many answers are correct.

Which of the following are direct benefits of the "*leaders speak last*" skillset? Choose all that apply

[ ] Demonstrate who the program leader is
[ ] Building a strong trust value
[ ] Increase focus on urgency
[ ] Creating and including diverse views

Which attributes showcase a fitting program roadmap?
Choose all that apply

[ ] Addresses performance metrics
[ ] Aligns with strategic vision
[ ] Focus on activities and tasks
[ ] Easy to follow visual across horizons

What is uniquely powerful about a well-developed program roadmap?

( ) Outcomes are visual and illustrate commitment
( ) Does not require product owners commitment
( ) Used even if there is no common program language
( ) Used to build program management maturity

3.5 POWERFUL STORYTELLING

Great leaders are excellent storytellers. Who does not like a good story?

We are affected by stories both mentally and emotionally. There is something unique about the power of stories. Program managers should do this well to connect the dots across stakeholders and processes. It is really almost a critical capability to have! If you look at the critical skills for the future, this is one that we need to continue to develop. Successful prioritization and communication excellence hinge on stories' use. Roadmaps are also an excellent example of telling stories.

Key Learnings

- Use PMBOK Guide development story as a backdrop for managing a profession's changes – shows how this happened over time
- Use NPR guidance for what a story flow should look like – an example of why there should be a flow to your program story
- Explore a few of agile ceremonies and their strength in storytelling transparency – the continual access across these ceremonies
- Creativity of connected program storytelling such as via use of a program canvas – very intentional in connecting the different elements of the program with another visual approach

3.5.1 The PM Profession Story

- Figure 3.11 depicts the journey of the PMI project management standards story from 1996 through the most current versions
- Maturing from process and knowledge focus to the principles and domains
- Moving project and program management to a more hybrid practice
- The program management world shifts from governance and gates to behaviors, such as we what talked about like, trust, discipline, teamwork, and other critical stakeholders relating attributes
- It is important for us to always describe the why and how then we got here
- A well communicated story, keeps us, and our program teams connected when we address how a program witnesses and creates shifts

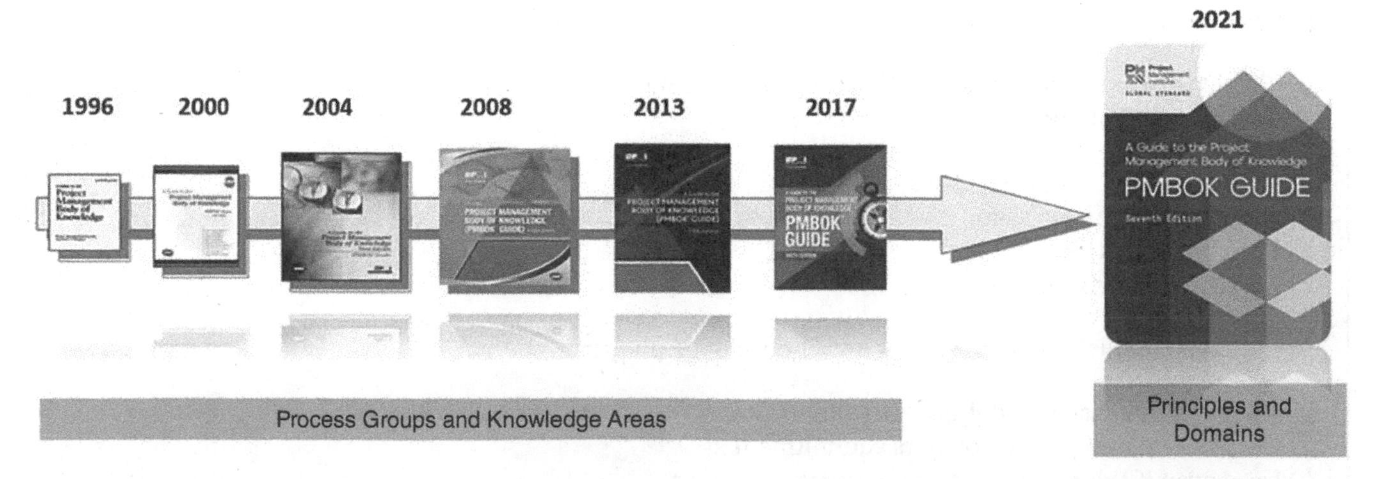

FIGURE 3.11 PM Profession Story Sample

- Maturity path
- The key principles shift
- The story linkages

3.5.2 The Storytelling Flow

NPR[2] and similar shared story taking tactics are useful. The flow is very important from the first story opening, just like in the movies: "you had me at hello." It is of note that attention span and attention currency are very critical to the future success of programs and especially with next generational changes.

- Beginnings are very important and match program success
- Builds the team empathy – if you don't build it, you lose your stakeholders
- Like a river – remember the cascading flow
- Clear call to action – team meetings should not take place unless you are leaving with a crystal-clear story around the actions to be taken next

> **TIP**
>
> Program managers garner tremendous value from practicing the principles of the right story flow.

3.5.3 Agile Ceremonies Role

Figure 3.12 summarizes the ways to use the agile practices to shift from the vague world to high transparency and bringing critical dialogues to the mix. Backlog prioritization is a good example of a strategic encounter with the product owner. Iteration planning has a clear focus on telling the story of what is coming next. Iteration review is a way to strengthen the future iterations and creating more clarity going forward. Continuous improvement is critical and this is something that great stories and high-quality interactions create.

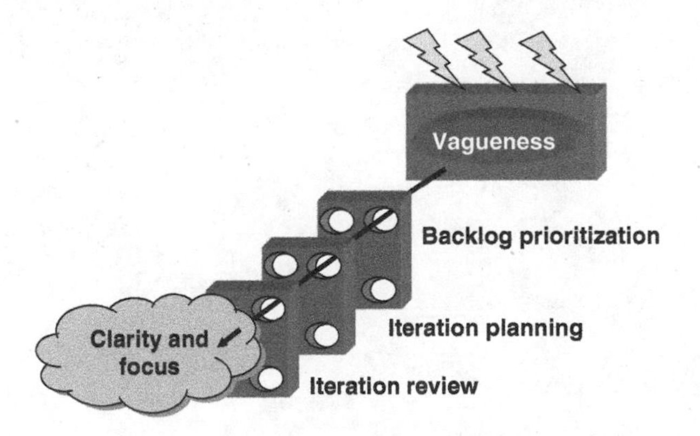

FIGURE 3.12 Communicating Across Agile Ceremonies

[2] NPR Storytelling Link: https://www.youtube.com/watch?v=tiX_WNdJu6w

Reminders

- Visual
- Links to strategic targets
- Focusses on value
- Illustrates change management
- Crystalizes strategic pillars

3.5.4 Program Canvas

A program canvas, like the tiles in the picture in Figure 3.13, is a nice organized template for showing critical pieces of program information. This makes the program elements visible to the stakeholders. It also forces integration naturally. You are able to highlight a multitude of the areas that you are tracking and keeping an eye on. This also brings the balance back to value. A program canvas acts almost like the basis for a big dashboard or an information radiator that brings the attention for what matters most for your program.

Reminders

- Integrates elements of program story
- Highlights balanced mix between strategic and value-based
- Allows program manager to play conductor role – the program manager is ultimately more successful when he/she has the connected program picture, as in the canvas

> **TIP**
>
> Program canvas is an example to go back to the core of integrating our focus on enhanced and connected planning.

FIGURE 3.13 Canvas Value
Source: Pixaline / 499 images.

Review Questions

Parentheses ( ) are used for Multiple Choice, when one answer is correct. Brackets [ ] are used for Multiple Answer, when many answers are correct.

_____ is a critical outcome of structuring the program story flow well.

( ) Having more fun.
( ) Building empathy.
( ) Noise.
( ) Focus on issues.

Which ceremonies contribute to creating stronger program team's focus? Choose all that apply

[ ] Detailed planning exercises.
[ ] Iteration reviews.
[ ] Retrospectives.
[ ] Iteration planning sessions.

What is most valuable in developing a program canvas?

( ) Integrating the program story.
( ) Presents a nice picture.
( ) High focus on activities.
( ) Only address resources.

SECTION II

APPLYING POWER SKILLS AND DIGITAL ENABLERS TO CREATE CONTINUAL CHANGE

SECTION OVERVIEW

This section describes the program management skills required for success in the future of work. It provides an opportunity to invest your time and energy in exploring what it takes to stretch your leadership capabilities further, while you uncover critical principles of program management. In the next decade, there is an expected critical balance between power/human skills and the digital enablers that are expediting the rate of change and expanding programs' impact on creating value to their stakeholders.

As a leader and a program manager, you will get the opportunity to learn and understand operating as a changemaker to align the program team members to the North Star of a given organizational transformation journey. You will learn a multitude of healthy "change culture attributes." There will be examples and case studies that get you to deeply understand the importance and mechanics of analyzing and engaging stakeholders, while appreciating the human side of change. An excellence model correlation will be made to the targeted future program managers' power skills.

For program managers to develop the right fitting future skills, they will need to fully immerse yourself in learning experiences. You have access to a wide variety of learning instruments and your success will depend on how you take the time to step back and block the time to think about the practices that I share with you and then personalize that to your specific learning and work needs. Your future role as the change maker will greatly depend on your ability to harness the balance between human skills and the future expected digital fluency. This is both a learning and a mindset shift journey. Program managers are expected to lead the world's most meaningful programs while we create continual change!

SECTION LEARNINGS

- The importance of the power skills for the future of work and how program managers need to continue to harness these.
- How to build the critical muscle to balance between power/human skills and the digital enablers?

Program Management: Going Beyond Project Management to Enable Value-Driven Change, First Edition. Al Zeitoun.

- Investing in the strategies and transformation steps, organizations need to take in order to be successful in the mindset shift journey required for the highly connected program teams of the future.
- Excellence in governance across organizations, driven by a program culture, critical questions to ask to ensure alignment is achieved, and the alliances needed for creating change.

KEY WORDS

- Culture Map
- Agility
- Power Skills
- Mindset
- Future of work
- Change making

CHAPTER 4

Change Making

This chapter is focused on the understanding of change making and the role of the change maker. It is intended to highlight the value of change in driving the future of business. This couples with a deep dive into the attributes for a healthy change culture. Examples and cases are used to exemplify sources of program complexity and how to tackle them. Program managers can benefit from the host of change capabilities that future program managers would need to develop as the change makers. Emphasis is also placed on the creation of stories that inspire successful change.

Key Learnings

- Review future businesses competing sustainability demands (what's happening and why sustainability is important in the work we do)
- Exploring the change-making capabilities to innovate the future organization (what is required in innovating the future organization)
- Using an example of the complexity in managing change – a Mega UAE energy program (we all face it and it is increasing with digitization, stakeholders' complexity, etc., and we will use a UAE example to help us understand dealing with complexity)

4.1 THE FUTURE OF BUSINESS

Businesses have been maturing in their understating of programs and projects, and how these have become a clear vehicle for transformation and growth. Major strategic agendas for organizations are executed via a connected portfolio that allows organizations to create the changes necessary for their key stakeholders and shareholders.

Program Management: Going Beyond Project Management to Enable Value-Driven Change, First Edition. Al Zeitoun.

4.1.1 Future Competing Sustainability Demands

Figure 4.1 is a reminder of the connected universe we find ourselves in. Sustaining the future for the next generations, while dealing with the competing demands of the environment, resources, demographics, and political unrests, is of utmost importance to organizations' agendas.

Change is the only constant. Making change not only happen but also stick is critical for sustaining the impact of programs and projects. Organizations and teams would need to consider a multitude of topics to handle these demands.

Strategic refocus (the make-up of our investments change):

- Country-level strategies
- Organization's strategic agenda (ultimately cascading to the program manager)

Initiatives' alignment (alignment between strategy and programs needs to be achieved):

- Integration criticality (need to strengthen this integration muscle)
- Value driver (benefits realization and the importance of bringing that across and putting that value front and center such as by using roadmaps and designing them around value)

Change Implications

- New set of behaviors (how we can get the program team to realize change is a must, getting into the mode of cherishing change attributes and aligning better with customers)
- New ways of working (how we collaborate, communicate, and interact while increasing the aptitude for handling changes)

FIGURE 4.1 The Sustainable Future
Source: geralt / 25607 images.

4.1.2 Change-making Capabilities

Figure 4.2 reflects the storytelling component that is at the heart of driving change culture. A set of topics are considered important for the future of business as change-making capabilities are being developed:

- Resilience (as evident in handling the world changes)
- Adaptability (needed for the changing ways of working)
- Cascading focus (strategic shift to aligning and having critical dialogues)
- Storytelling (need to be continuously developed, including focus, pace, emphasis, and how the story connects the dots for the program and its teams)
- Brainstorming (super important in the future of work and for creating customer-centricity)
- Design thinking (helps with iterating through the pain points and designing solutions around better customer requirements)

FIGURE 4.2 Making Change Happen
Source: darkside-550 / 226 images.

4.1.3 Complexity in Managing Change

Complexity is never going to go away. Figure 4.3 highlights the message that change complexity centers around the people involved in the change process. In managing change, a number of complexity factors come into the mix in successful implementation of the change. Most of that complexity relate to stakeholders, technology, and communications.

A UAE Energy Journey

- First nuclear power plant in the UAE and region's multi-unit (something new and hasn't been done before)
- Korea Electric Power Corp. (KEPCO), APR1400 is a next-generation nuclear power plant (shadow plant was being constructed in South Korea and learning from that is studied and replicating the learning on the grounds in Abu Dhabi, connecting the dots to inclusion of new technology while creating a new safety culture)

FIGURE 4.3 Complexity Management
Source: geralt / 25607 images.

- Most diverse multicultural nuclear program in the world, with over 40 nationalities involved (different way of thinking, communicating, and collaborating, and thus how can the PM adapt and use the learning to create the highest impact and connectedness across the variety of stakeholders)

Same qualities are needed in simple programs that for sure will have some source of complexity or another. We need to have strategic tactical solutions to deal with programs' complexity.

Review Questions

Parentheses ( ) are used for Multiple Choice, when one answer is correct. Brackets [ ] are used for Multiple Answer, when many answers are correct.

_____ is a critical shift in how future organizations excel in meeting their strategic sustainability agendas

( ) New set of behaviors
( ) Value driver
( ) Country-level strategies
( ) The integration criticality

Which of the following change-making competencies directly contributes to your ability to handle the programs' VUCA environment?

Choose all that apply

[ ] Cascading focus
[ ] Resilience
[ ] Storytelling
[ ] Adaptability

Which is of the following example of a by-product benefit of the UAE energy program?

( ) Illustrating and learning from a comprehensive diversity case
( ) Winning excellence awards
( ) Happy citizens
( ) Being the first multi-unit plant in the Arab region

4.2 CHANGE CULTURE

The tone from the top sets the culture of the organization and the foundation of whether or not the change muscle will grow and be effective in handling the achievement of challenging outcomes. Healthy change culture means that there is the right understanding of the importance of the change mission. It means that there is the right investments and level of support for the initiatives that are expected to take us on that journey of change.

Key Learnings

- The worst scenario – no healthy change culture (how to alter that?)
- Exploring cultural attributes to enable change (what do I need to be supportive of?)
- Linking to Excellence and exploring the EFQM Model (so many attributes to reflect on around program excellence and creating the right excellence culture)
- Common reasons for impact of culture on programs failure (why is this an important investment of my energy and time?)

4.2.1 Implications of an Unhealthy Change Culture

Possible Implications:

- Management and/or key stakeholders could impose unrealistic expectations (not having the right dialogue or the right sensing going on)
- Ineffective programs governance (really a bad sign as we need to have the ability to be authorized as PMs to make the right timely decisions)
- Resistance to acceptance of new ideas and innovations (massive impact on program success, need to work with the stakeholders, and create alliances)
- Constantly changing priorities (a classic implication, teams have no idea why things are changing, they lose trust, and becomes negative implication on program teams' ability to continue to push forward)

4.2.2 Cultural Attributes to Enable Change

Figure 4.4 is a reminder of the criticality of choices leaders make in order to set up the right healthy culture, built on a set of key attributes. **Vulnerability example:** the work from Brené Brown we addressed before is of great quality to cut through the noise and focus better

- Cut through noise and design a joint view of benefits, make mistakes, and learn fast

FIGURE 4.4 Cultural Choices
Source: Pixource / 95 images.

Humbleness Example (Notion of Service)

- Roll up your sleeves (balancing when to jump in), stay back when suitable, and use the power of acknowledgement timely (giving credit to the ones contributing the highest impact on program success)

Colliding Philosophies Example

- Listen for a change (deepen listening capabilities), build critical bridges (networking and alliances), and have tough conversations often to grow your EQ (goes a very long way in terms of supporting the right culture necessary for creating change)

4.2.3 Excellence and Exploring the EFQM Model

As shown in figure 4.5, cultural-related attributes, such as being clear on why we exist, the agility in the program leader, excellence and creativity and innovation go hand in hand, by nature programs bring diversity to the mix with contradictory approaches and rich experiences, also create a culture that includes, innovates, and sustains following agendas like the United Nations Sustainable Development Goals (UN SDG).

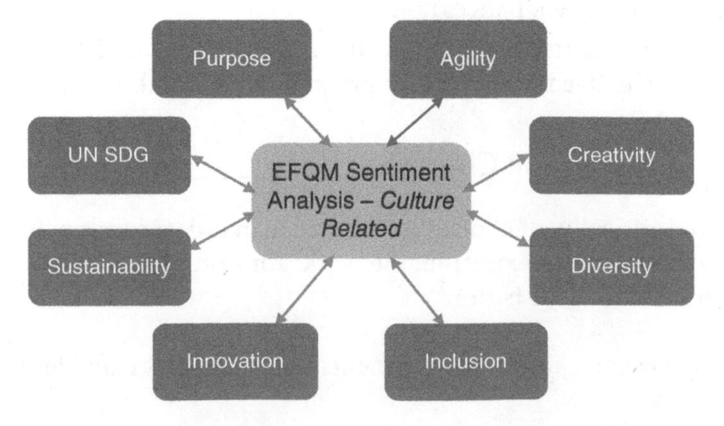

FIGURE 4.5 EFQM Cultural Elements
Source: Adaption from the European Foundation for Quality Management (EFQM).

4.2.4 Impact of Culture on Programs Failure

- Cultures that focus mainly on command and control of hierarchy do not support program team's flow (will not exist in the future)
- Executives who are rewarded for meeting budgets and deadlines may not be focused on programs benefits (completely shooting the leaders' ability to focus on value, so we need balance and not win the battle and lose the war)
- Missing collaboration between program managers and resource managers leading to failure of matrixed way of working (classic aspect of cultures that have to change and see the benefits that the various business areas' teams are bringing to the mix and streamlining the learning in their areas, and have consistency in delivery to mature our cultures)

Review Questions

Parentheses ( ) are used for Multiple Choice, when one answer is correct. Brackets [ ] are used for Multiple Answer, when many answers are correct.

_____ is a critical consequence of an unhealthy culture that affects the program manager's decision-making capacity.

( ) Collaboration
( ) Ineffective program governance
( ) Resistance to new ideas
( ) Dynamic management support

Which of the following EFQM sentiment analysis addresses cultural excellence dimensions? Choose all that apply

[ ] Sharing Ideas
[ ] Agility
[ ] Trust
[ ] Sustainability

Which is of the following example of metric that affects weakened focus on benefits?

( ) Illustrating leadership by example
( ) Rewarding executives mainly for meeting budgets and deadlines
( ) Sharing information freely
( ) Being the first to say no to change

4.3 CHANGE MATTERS

Key Learnings

- Is the Program Manager suited to drive change?
- Defining changing the business versus running the business
- Why changing the business is fun and impactful?
- Strategic change and how it maps to clear organizational vision (the importance of why the strategy contributes to the strategic efforts that are taking place in the form of programs and projects)

4.3.1 Program Managers and Driving Change

Figure 4.6 brings together some key building blocks for driving change. These have been addressed across various chapters of the book. The ***initiatives designed to create change*** build block are meant to highlight the importance of purpose and of having a joint view of success for the program among its stakeholders. ***The ability to work across organizational boundaries*** building block remains one of the most highly targeted organizational changes in support of cross business initiatives' success. This requires commitment and dedication of the executive team to drive the organizational culture and reward mechanisms in that direction. The ***conductor qualities*** building block that has been looked from many angles throughout the book, remains one of the most value-added qualities that program managers bring in their ability to drive connected change. ***Bringing trenches experiences*** building block is naturally quite valuable. This enables the leader to have real views and ideas that enable solving the process, technology, and people problems, typically associated with creating changes. Finally, the ***ability to see the business end-2-end*** is increasingly one of the qualities that differentiate leaders' ability to drive change. This means not only having the holistic view but also having a strong sense of the value stream of the program and its lifecycle. This is critical in the highly customer-centered and agile way of working we encounter now and into the next decade.

Talking about the program manager's personality attributes, and whether that leader is the right one, the following list of qualities and skills come to the surface:

Handling changes, working across boundaries internal and external, being the conductor, bringing the qualities for integration, and bringing real experiences especially around working with people and process, see the business holistically, getting on the balcony for distance, and doing whatever it takes for you to provide the kind of efforts needed to handle the change expectations of these programs.

> **TIP**
>
> Focus on developing the five building blocks that increase your ability to drive changes that are strategically meaningful to the program's stakeholders.

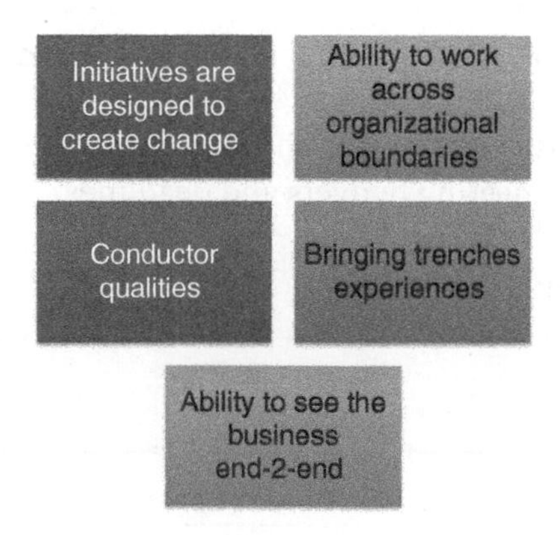

FIGURE 4.6 Driving Change

4.3.2 Changing the Business versus Running the Business

Organizations are continuously becoming more creative in their business portfolio mix and are shifting:

Source: mohamed_hassan / 6079 images.

- Strategic versus tactical focus (the more we realize that programs are strategic vehicles, the more that shift increases)
- Differences in thinking muscles use (programs require us to be futuristic in thinking)
- Impact on programs value (how are the programs creating the continual impact on achieving value?)
- Approach to stakeholders' engagement (strategic, future orientation, dealing with risks, and changing minds and ideas)

> **TIP**
>
> Organizations are encountering big shifts that require consciousness in understanding the changing of the business capabilities.

4.3.3 Change the Business Is Fun and Impactful

Program managers are in a position that gives them ability to grow and learn ***future qualities***:

- Flexibility and adaptability
- Confidence, persuasiveness, verbal fluency (enhancements in your communication style)
- Ambition, activity, and forcefulness (to deal with resistance)
- Effectiveness as a communicator and integrator (great to complement your changing the business success)
- Broad scope of personal interests
- Poise, enthusiasm, imagination, and spontaneity (great opportunity to truly get your learning sponges to grow)

See the direct impact on organizations and individuals:

- Strategically mapping the work of the program to business results (most exciting aspect about programs, seeing the connecting points, and cascading that across your program team)

4.3.4 Strategic Change and Clear Organizational Vision

Making change happen requires a high degree of vision alignment across a vast group of stakeholders. Figure 4.7 highlights some of the key contributors to that clear organizational vison. Many elements contribute to having this clean lens. The mix of operational excellence, connectedness to the customer, understanding of the external environment and internal focus of the executives in investing time and resources in what matters, come together to achieve such an important connected vision.

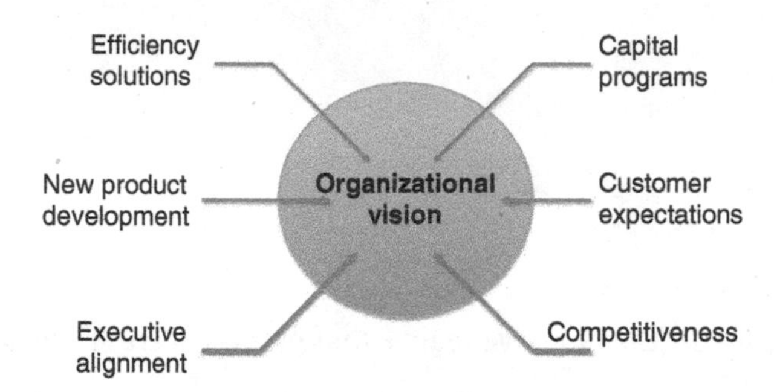

FIGURE 4.7 Organizational Vision Contributors

Many of these different types of programs could have a place in the strategic agenda of the organization and have different set of expectations, complexities, and stakeholders. Thus, we need to develop the sensitivity to deal with this variety and have the leadership qualities to support our success.

Review Questions

Parentheses ( ) are used for Multiple Choice, when one answer is correct. Brackets [ ] are used for Multiple Answer, when many answers are correct.

_____ is an example of what qualifies a program manager most to drive change.

( ) Ability to work independently
( ) Bringing trenches experience
( ) Deep technical expertise
( ) Saying no often

Which of the following are positive and fun elements of programs that are focused on changing the business?

Choose all that apply

[ ] Showing who has the most power
[ ] Confidence and adaptability
[ ] Drive the change in a direction chosen by the program manager
[ ] Witnessing the impact, the program outcomes create for the customers

Which of the following are initiatives example that directly support achieving organizational vision?

( ) New Product Development
( ) Running the new ERP implementation
( ) Happy stakeholders
( ) Using less resources to run the business

4.4 THE INSPIRING PROGRAM STORIES

Key Learnings

- Positioning Program Manager capabilities with an eye on connected storytelling role (need to dedicate the time to do it)
- The Program Team Charter example as key to team's dynamics story (this is designed for the teams)
- Tailoring change stories to inspire successful change will be demonstrated (the whole journey of driving programs to achieving change while using power skills, hinges on our ability to tailor around the constraints, stakeholders, and other criteria)

4.4.1 Program Manager and Storytelling

Figure 4.8 highlights not only the ingredients to strengthen your storytelling abilities but also the need for a dynamic and changing nature of stories across the program lifecycle.

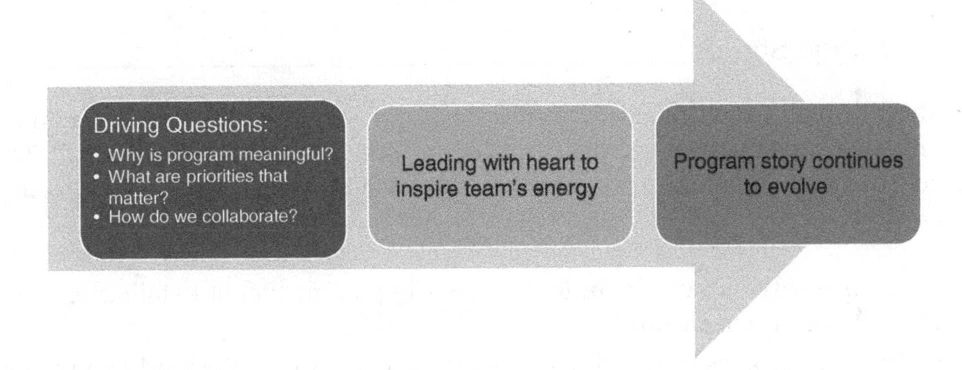

FIGURE 4.8 Stories for Leading

Reminders

- Your ability to communicate
- Finding exact questions that help you focus
- Questions across delivery approaches
- The right mechanism to connect the story
- This connection between the head, heart, and hand so the team could see your energy
- Not only one story as we need to consider progressive elaboration across the phases of the program, the changing maturity, and stakeholders

4.4.2 Program Team Charter

The program ***team charter*** is a key fun tool that helps in driving the secret sauce that connects the program team and enable them to drive change outcomes. We typically don't take the time to think through this or capture that in some form of a document. This team charter is important for driving the right interactions, connect the team members to why the team exists. This charter has key elements as shown in Figure 4.9. It is typically a short template of a couple of pages that addresses the expectations including how to handle conflicts, team performance, creates the join commitment, and clarifies other accountability points.

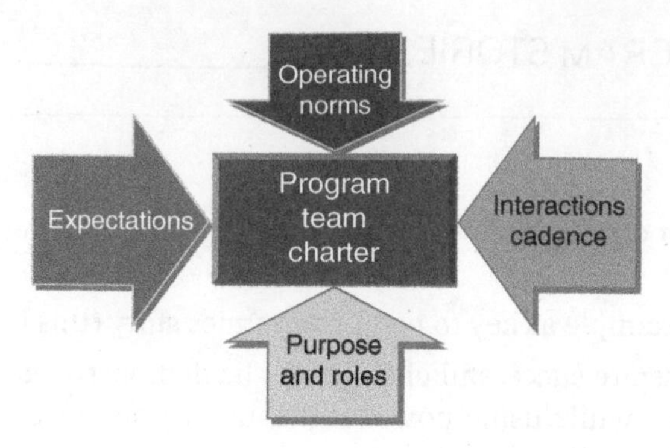

FIGURE 4.9 Team Charter Connections

> **TIP**
>
> Team charters help the program manager in driving change. You got to get this agreement vehicle intact and socialized across your team.

4.4.3 Tailoring Change Stories

Figure 4.10 highlights the need for continual adaptability in shaping the change stories. Customization and being sensible to the changes that the program manager encounters, is becoming a must-have skill. The following list summarizes multiple areas the program managers should consider in tailoring the change stories they use to drive change forward and connect their program stakeholders.

- Context (priority, maturity, success definition) . . . very important anchor to tailoring, any of these points when they change, we have to customize
- Culture and virtual aspects (as in working or residing across parts of the world and the developing way of working)
- Stakeholders' state of mind (including attitude, clarity of roles and responsibilities, and thus you need the story to bring them back on track, have a flow to the story and a clear call to action to the stakeholders to get them back to focus)
- Program delivery approach (between waterfall versus the fluid and adaptive delivery, so a need to customize way of working and ceremonies)

FIGURE 4.10 Tailoring Plans
Source: geralt / 25607 images.

Review Questions

Parentheses ( ) are used for Multiple Choice, when one answer is correct. Brackets [ ] are used for Multiple Answer, when many answers are correct.

_____ is an example of a driving question that supports inspiring the program team.

( ) What meetings should the team attend?
( ) What are the priorities that matter?
( ) When do we freeze the program scope?
( ) How much funding is the program expected to receive?

Which of the following are advantages of developing and using a program team charter? Choose all that apply

[ ] Roles and responsibilities
[ ] Ground rules
[ ] Showing rank
[ ] Frequency of meetings

Which of the following is an example of tailoring a program story to address the way by which a Korean colleague prefers to receive feedback?

( ) All team members should equally be able to understand the program story
( ) Tailoring for state of mind
( ) Tailoring for culture
( ) Tailoring for delivery approach

4.5 TRANSFORMATION QUALITIES

Key Learnings

- Balancing people, process, and technology to drive transformation success or else (balance is an important reminder for what is needed in transformation)
- Volatility, Uncertainty, Complexity, and Ambiguity (VUCA) is here to stay (this is where we are)
- Using SAFe model to manage scaling transformation programs (one of many models that could help us with scaling, so find the right one)
- Identifying role of establishing continual program connection between head and heart to succeed in achieving execution outcomes (repeatable important point for the success in driving transformative impact)

4.5.1 Balancing People, Process, and Technology

Any time teams are on the hunt to create change, these three elements of people, process, and technology have got to be addressed and balanced. Programs that focus on one at the expense of the others could suffer and miss the boat on what matters most in terms of the ultimate program benefits targeted.

- Humans will have to develop proper mindset and characteristics that are a must in tomorrow's innovative organizations (need to insist on having a different mindset such as the aptitude for learning).
- Innovation and technology ingredients cover autonomy/alignment balance, safe culture, collaborative strength, time to think, and seamless organizational flow (making sure that we will have the focus on the intentionality in driving these behaviors).
- Leadership of future requires a responsible mindset: Humans who are not stuck (having leaders who are not stuck and have a strong sense of ownership) and true connected players who are organizational performance focused.

4.5.2 VUCA Is Here to Stay

Source: Tumisu / 1245 images.

The elements of Volatility, Uncertainty, Complexity, and Ambiguity of VUCA will continue to remain and evolve in the future of work. As highlighted in Figure 4.11 program leaders show organizations the way for how to best handle VUCA and bring the programs to calmer shores, thus increasing the changes of their success.

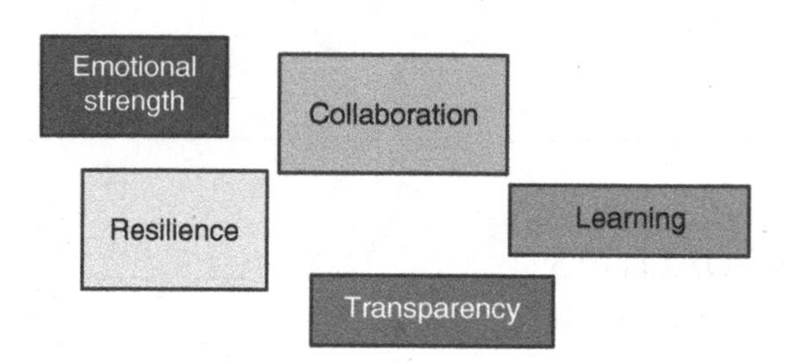

FIGURE 4.11 VUCA Handling Ingredients

Program managers should think about ***stretching*** ourselves, intensifying ***collaboration*** and making it intentional, creating an environment that supports developing ***transformative leadership***, and practicing ***transparency and resiliency*** affecting how the teams succeed in dealing with the VUCA environment surrounding our programs and team.

4.5.3 SAFe[1] and Scaling Transformation

Reminders

- Collaboration across SAFe layers
- Role of governance in scaling (the ability to enhance and focus on the clarity of governance)
- Engagement and integration around value (massive engaged and connected entities of stakeholders)
- Faster time to market (including the speed for creating impact)

[1]SAFe: https://v5.scaledagileframework.com/safe-lean-agile-principles.

4.5.4 The Head and the Heart's Continual Connection

To achieve ***Strategy Execution Excellence***, transformative leaders connect the dots and put excellence front and center in how they operate and the way of working they develop for their program teams:

- Clear transformation of mental focus
- Aligned hearts around values, illustrated in figure 4.12 (this requires storytelling, roadmaps, and adapting and dealing with the punches)
- Adaptable execution

> **TIP**
>
> Leaders understand the critical connection between the head and the heart. The head drives pragmatic focus and the heart provides the energy for change.

FIGURE 4.12 Connecting the Head and the Heart
Source: geralt / 25607 images.

Review Questions

Parentheses ( ) are used for Multiple Choice, when one answer is correct. Brackets [ ] are used for Multiple Answer, when many answers are correct.

_____ is a critical ingredient that allows people the innovation required for achieving transformation goals.

( ) Stubborn mindset
( ) Safe culture
( ) Focus on action
( ) Detailed planning

What are the future skillsets that help program teams to deal with the continual presence of the VUCA environment?

Choose all that apply

[ ] Resisting change
[ ] Resilience
[ ] Baselining
[ ] Emotional strength

Why is it critical that the head, heart, and hand are in alignment to achieve transformation outcomes?

( ) Illustrating hierarchy in the organization
( ) Illustrates the team's ability to execute what matters most
( ) Makes stakeholders happy
( ) Being human

CHAPTER 5

Effective Engaging

This chapter focuses on understanding the importance of engaging stakeholders and program teams around the program's direction and the expected integrated set of steps that have to take place in a coordinated fashion toward the achievement of program's outcomes. Programs' success is directly correlated to effective stakeholders' engaging. Understanding the landscape of the extended groups of program stakeholders is critical.

Program managers should reflect on examples for how tiers of stakeholders can be categorized and designed. They should learn how to analyze and engage stakeholders. This learning becomes the foundation for exploring the design of the fitting engagement strategies. In addition, and due to the very fluid nature of programs, one should develop an appreciation for sensing the engagement changes necessary across the program life cycle.

Key Learnings

- Resilience capabilities and the Elon Musk example (what examples do we see in this case in leading massive programs?)
- How engagement strategies vary across the program life cycle? Adaptability component
- Understanding organizational viruses that impede engagement (creating the engaged community of stakeholders is a must)
- Battling organizational viruses with proper engagement antidotes (how to deal with the viruses?)

5.1 ADAPTING ACROSS THE LIFE CYCLE

One of the critical capabilities that support the program manager's success in adapting to the changing needs for engaging stakeholders is resilience.

Musk Resilience Example

- Have a vision and always dream big (something that program managers should have that demeanor and mindset and connect others to it)

- Execute your ideas before it's too late (timeliness is a difference maker while experimenting)
- Be persistent (very important attribute like in Musk' case)
- Think about the bigger picture (always connect the dots)
- Look out for constructive criticism (feedback is it!)
- Consider failure an option and take calculative risk (not an end, strengthening risk aptitude)

5.1.1 Engagement Strategies Across Program Life Cycle

The continual improvement approach to engaging stakeholders, as highlighted in Figure 5.1, is an indication that there is a need to continually assess what is working and what is not and quickly adjust the engagement strategies accordingly.

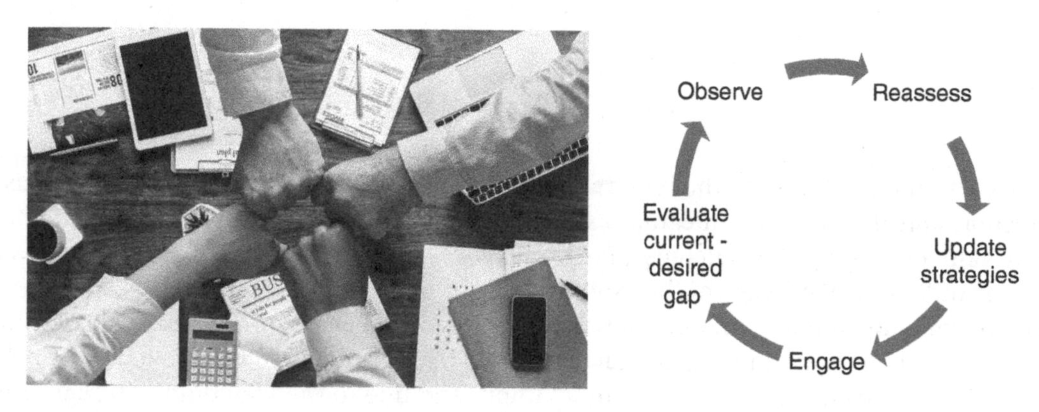

FIGURE 5.1 Engaging over the Life Cycle
Source: mohamed_hassan / 6079 images.

There are multiple causes why the strategies have to remain fluid:

- Stakeholders change across the phases (need to adapt and we could use the phases as an opportunity for that, some of the change is natural)
- The role and influence they play change (following the circle shows the continuity of the steps and how to close the gaps between where stakeholders are at and where we want them to be, like we have in a classic continuous improvement cycle)
- Level of interest changes

5.1.2 Organizational Viruses

Program managers should identify and handle organizational viruses. These are cultural, way of working patterns, and behaviors, that organizations develop over time.

Examples of These Viruses Include:

- Unhealthy politics
- Lack of accountability (missing ownership)
- Resistance and insecurity about program-created changes (this notion of needing to use power skills to manage resistance)
- Poor sponsorship (right expectations and understanding of their role)
- Gaps in engagement infrastructure (platforms, the fitting of engagement components for working across cultures)
- Not invented here syndrome (really makes the program manager (PM) get concerned whether the right attitude and commitment exist and thus tackle the creation of coalitions to address this soonest)

5.1.3 Proper Engagement Antidotes

Figure 5.2 shows a set of antidotes that would help address engagement challenges and potentially kill some of the organizational viruses highlighted above. Investing in ***transparency***, visibly rewarding a culture of ***ownership***, ***storytelling***, taking the player role in finding ways to communicate to executives, use ***technology*** to directly drive programs success, and finding the fitting ***leadership styles*** mix to deal with the viruses.

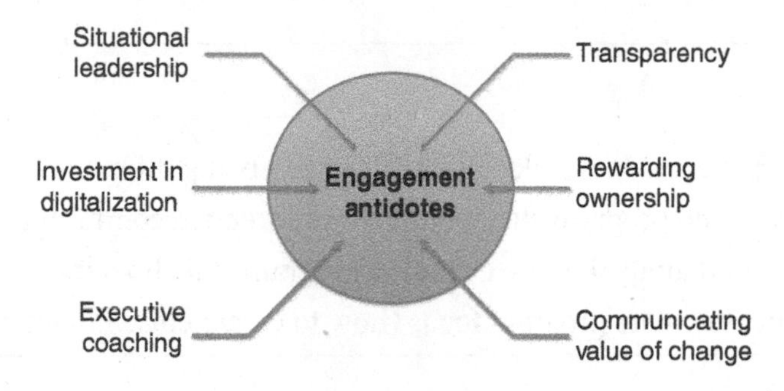

FIGURE 5.2 Engagement Antidotes

Review Questions

Parentheses ( ) are used for Multiple Choice, when one answer is correct. Brackets [ ] are used for Multiple Answer, when many answers are correct.

_____ is an example of resilience qualities a PM could learn from Elon Musk.

() Risk aversion
() Execute fast and take risks
() Stay mainly focused on the program details
() Disregard criticism

Which of the following are examples of viruses that would impede proper stakeholders engagement? Choose all that apply

[ ] Digital expertise
[ ] Lack of accountability
[ ] Staying away from politics
[ ] Not invented here syndrome

Which is of the following antidotes is helpful in dealing with the resisting change organizational virus?

() Investing in office equipment
() Hierarchical leadership
() Multiple points of accountability
() Communicating value of change

5.2 PROGRAM STAKEHOLDERS

Mahatma Gandhi said, “You must be the change you wish to see in the world.” To get the change expected of a give program achieved, a village of stakeholders need to align around that change and contribute to achieving the planned results. Leading by example is an increasingly important way of working in the future as there will be no room for any form of classic management and much more space for servant leadership and strong willingness to connect across groups of stakeholders.

Key Learnings

- Rethinking what program success looks like (open to success being different)
- Expanding the view of who the program stakeholders are (extended community)
- Technique for identifying and analyzing the extended program stakeholders
- Using tiers of stakeholders to create program focus (how to create engagement efficiencies?)

5.2.1 Rethinking Program Success

Figure 5.3 reminds the PM of the steps leading to continual clarity of what success looks like and the importance of rethinking success as necessary in order to better adapt to the program's changing ecosystem of stakeholders and other variables.

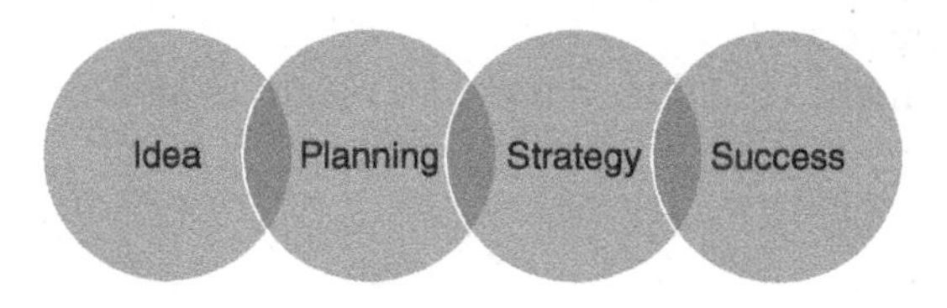

FIGURE 5.3 Rethinking Success
Source: geralt / 25607 images.

- Have all right idea questions been asked? (Even before the journey starts, do whatever we can to find out where the stakeholders stand)
- Are plans reflective of the program's complexity and include key inputs? (Plans need to be a contribution of the team and stakeholder's)
- Is there commitment to execution approach? (Team charter and other tools we addressed)
- Is there alignment on view of program success? (Rethink success and ensure that it focuses on achieving benefits and value, debate common views)

5.2.2 Expanding View of Program Stakeholders

Missing a key stakeholder could make all the difference in the likelihood of being successful in a program or a project. The program leaders are much better off expanding their views of who the stakeholders are to ensure a wider spectrum is captured and the right assessment and plans for engaging, when necessary, are in place. Key stakeholders are not just the customer, users, or the executive team. The range could be the wide body of the 3Is (Interested, Involved, or Impacted).

- Expanded stakeholders, include multiple entities and individuals with interest, involvement, or impact:
  - Sponsors
  - Regulators

- Team (sometimes multi teams)
- Community
- Regulators (also standards setting bodies)
- Many more (expand our views and keep them in balance)

5.2.3 Technique for Assessing the Extended Stakeholders

Culture across regions and within organizations plays a critical part of sensing and responding to stakeholders. This cultural understanding ties nicely to sensing and responding qualities needed as program leaders. The PM could assess the adjustments necessary to work with diverse stakeholders around the eight cultural elements highlighted in figure 5.4 by Erin Meyer.

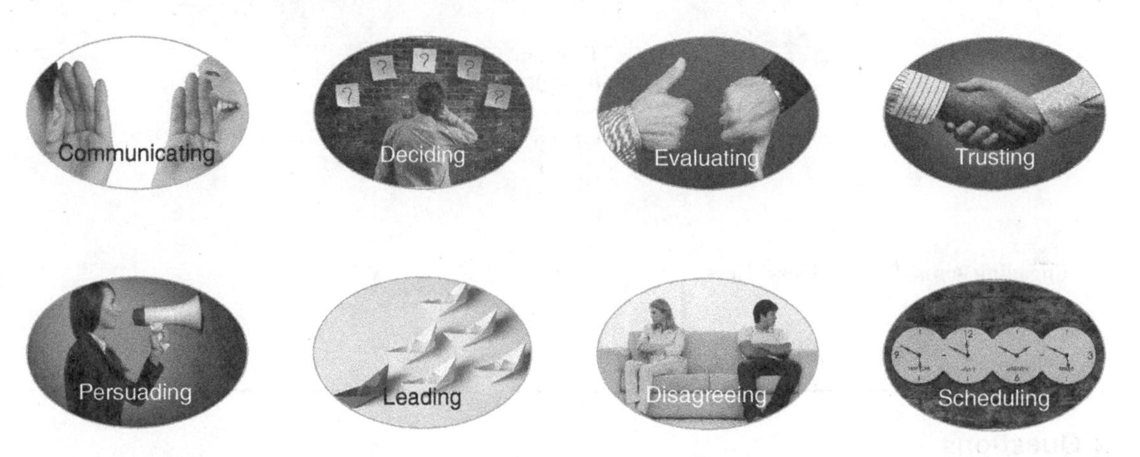

FIGURE 5.4 Expanding Cultural Understanding
Source: Faber Visum/Adobe Stock Photos; Brian Jackson / Adobe Stock; sharpshutter22 / Adobe Stock; yossarian6/Adobe Stock Photos; WavebreakmediaMicro/Adobe Stock Photos.

Each one of these eight elements could vary on a scale between regions and countries of the world. Program team members need to be respectful of these cultural elements' differences and include that in the design of their engagement strategies.

> **TIP**
>
> Organizational leaders understand the world of work is global. Understanding these eight elements enables the leader to excel in connecting to stakeholders.

5.2.4 Stakeholders Tiers and Creating Program Focus

Stakeholders' groups vary from the direct ones around the PM, to the extended entities, all the way to partners, customers, and others. The idea is to create some organized structure and achieve efficiencies in how to design strategies that work for groups of stakeholders, as reflected in the tiers shown in figure 5.5. Strategies need to be intentional and yet speak to them in a language that they can understand. This is not something done by one person and the program manager needs to rely on a core team around him/her to design tactics and engagement strategies. Continual exercise of adjusting strategies to get back to program focus is important and correlate well to program success.

- Prioritization of tiers based on multidimensions
- Efficiencies across engagement strategies
- Alignment driven by a core program team

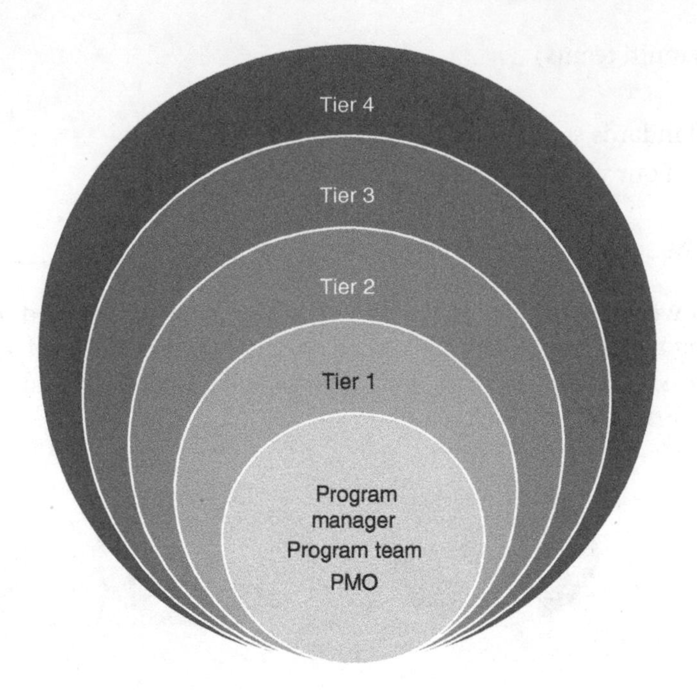

FIGURE 5.5 Engaging across Stakeholders' Tiers

Review Questions

Parentheses ( ) are used for Multiple Choice, when one answer is correct. Brackets [ ] are used for Multiple Answer, when many answers are correct.

_____ is a good example of what qualifies as program success.

( ) Having the best members on the program team
( ) Program plans are reflective of the program complexity and include key inputs
( ) Most creative ideas are chosen for the program
( ) Policing the program outcomes

Which of the following is a culture dimension that directly correlates to openness of feedback on an agile program team?

Choose all that apply

[ ] Deciding
[ ] Evaluating
[ ] Persuading
[ ] Communicating

Which of the following is an example of a challenge faced when dealing with tiers of stakeholders?

( ) Lack of alignment across stakeholders
( ) Using the right assessment tool
( ) Happy stakeholders
( ) Having only three tiers of stakeholders

5.3 ENGAGING STAKEHOLDERS

Key Learnings

- The worst scenario – not being engaged (could be a nightmare, so needs to be addressed fast)
- Linkages across program stakeholders and managing complexity
- Exploring the connection between proper engagement strategies and the 30% of transformations that succeed examples to understand engagement matrices (what is the secret formula that is missing? Many time it is related to the people side of the transformation)

5.3.1 Effects of Missing Engagement

- When do programs/projects fail? (Early or before they start)
- Could you do everything correctly and still fail? (It is not just about delivery or busy work as we addressed before)
- How well are we sensing our stakeholders? (Very important to improve ways of relating)
- Do we understand cultural diversity implications on engagement success? (We looked at the eight elements form the work of Erin Meyer and how this can help us adjust our engagement strategies)
- How agile do we remain against the continual organizational and project changes and fires? (So important to adapt and deal with business and political changes)

5.3.2 Stakeholders and Managing Complexity

Strategy is difficult. Executing strategy is complex. Figure 5.6 indicates an attempt in the picture to show the many models we could use to execute strategy, cascade vision, and mission into programs and projects outcomes.

Reminders

- Program complexity dominates current VUCA environment (openness to different models)
- Complexity has been tackled with a multitude of models to enable executing strategies via programs and projects (simplicity matters)
- Most models have a key stakeholders' engagement element (every model has an obvious and key stakeholders' engagement component)

TIP

Leaders understand the importance of finding the most fitting approach to use in aligning stakeholders around the expected strategy execution outcomes.

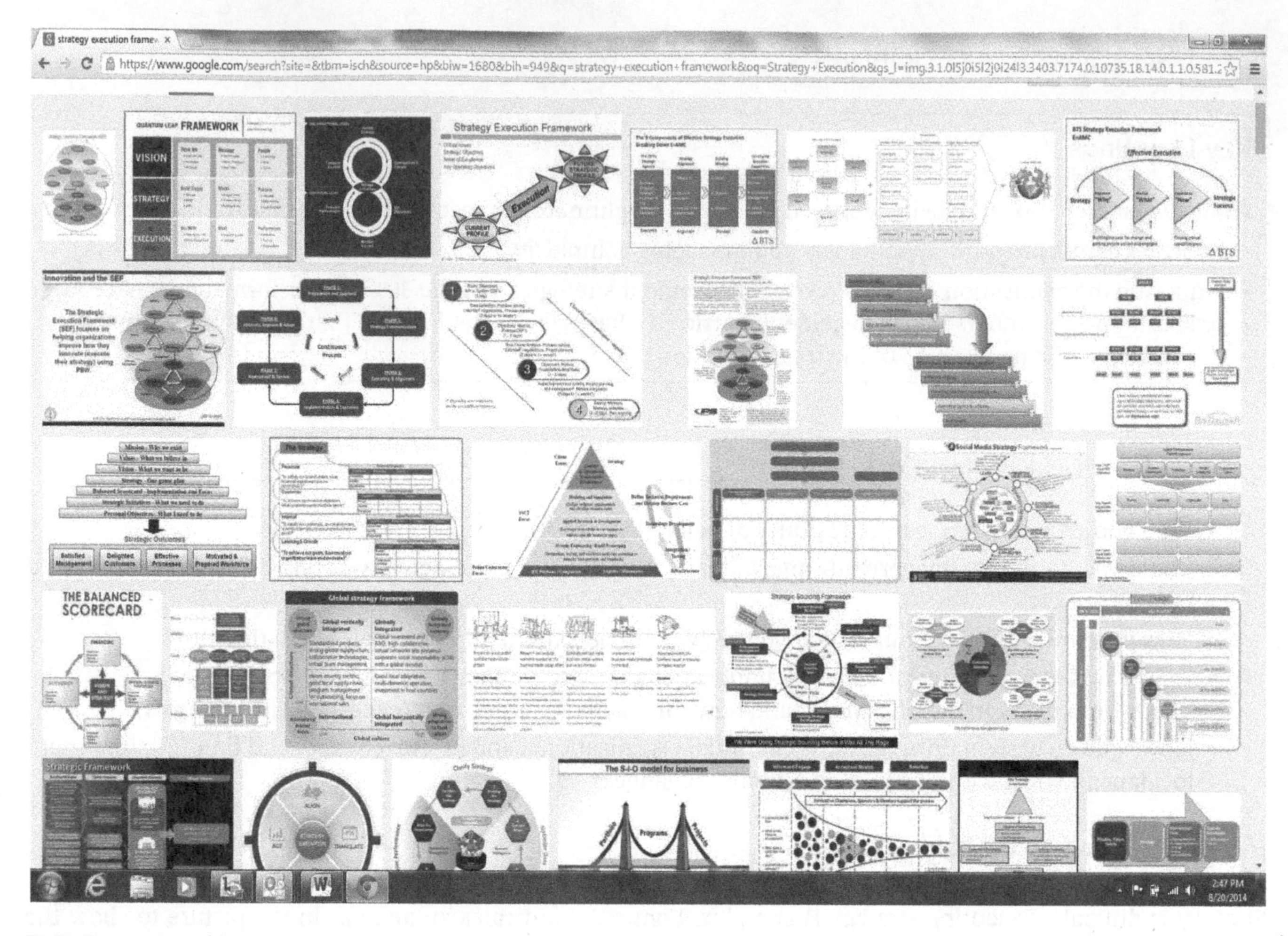

FIGURE 5.6 Tackling Complexity

5.3.3 Engagement Strategies and Transformations Success

Figure 5.7 shows the circle used previously to continuously update one's strategy to engage and achieve closure of any communication gaps.

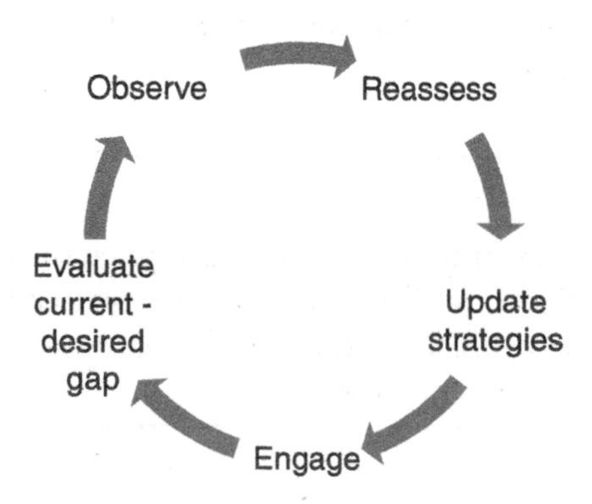

FIGURE 5.7 Continual Engaging

It is likely that the small 30% of transformation programs that succeed is good at

- Flexibility and adaptability
- Progressive elaboration around engaging stakeholders (100% critical)

- Creativity in filling the engagement void (need to try a few of these individually or collectively)
- Linking transformation to stakeholders' view of value
- Successful transformations are people-centric (it is not the process of technology alone, need this secret ingredient for transformation success)

Review Questions

Parentheses ( ) are used for Multiple Choice, when one answer is correct. Brackets [ ] are used for Multiple Answer, when many answers are correct.

_____ is an example of a situation where the engagement was the missing ingredient for program success.

( ) Asking the right questions timely
( ) You do everything correctly and the program still fails
( ) PM operated as a conductor
( ) Open feedback process was prevailing

Which of the following are signs where complexity around strategy execution is linked to proper engaging of stakeholders?
Choose all that apply

[ ] Increased investment in running the business work
[ ] Most strategy execution models have a critical stakeholder's dimension
[ ] Drive the change in a direction chosen by the program manager
[ ] The ambiguity associated with programs' requirements

Which is of the following is a contributor to transformation programs success?

( ) Sticking to one structured approach to running these programs
( ) Being technology-centric
( ) Linking transformation to stakeholders' view of value
( ) Creating a bassline engagement plan and keeping it

5.4 ENGAGEMENT STRATEGIES FOR FIT

Key Learnings

- Myth behind success and a pharmaceutical organization's case study (understand the need for customization)
- Using case study to highlight essential need of customization for proper engagement
- Multiple factors affecting the tailored engagement (bring together the points that PMI has added in its standards, PM should lead this tailoring role)

5.4.1 Myths Behind Success

This is such an important realization for the PM. Although the classic success drivers are possible, we need to rethink what the drivers for success are. The program leader could be shocked if the real needs are beyond the agreements or contracts. Leaders need to be vigilant in cooperating across the stakeholders' groups and continuously seeking and including their feedback.

This pharmaceutical case study of mistakenly thinking that this global organization was happy since they signed off on the deliverables checks the box on all the myths highlighted in Figure 5.8. In fact, true success was not there, and this was mostly attributed to the fact that proper engagement was not there.

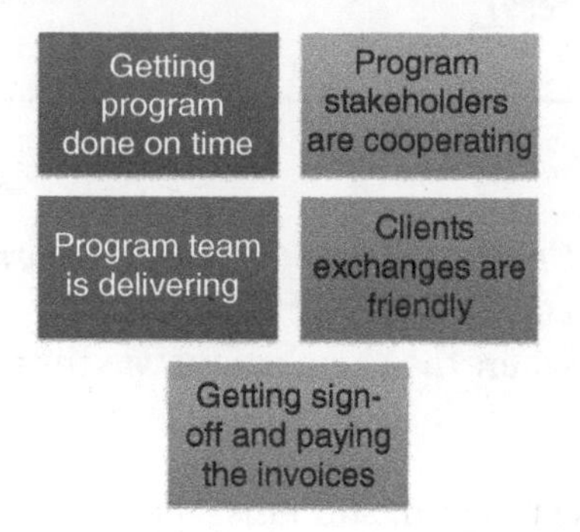

FIGURE 5.8 Success Myths

> **TIP**
>
> Program managers must comprehend the success myths. They have an accountability to properly align with key stakeholders on joint accurate success views.

Need for Proper Engagement

- In this pharmaceutical organization case study, a multitude of ***wrong assumptions*** were made and never got clarified: (we have to be very clear on what the assumptions are and how they are changing, and that stakeholders are linked to those changes)
  - Program SOW covers the complete stakeholders' expectations (nothing in this case was indicating that the organization was concerned about managing resistance of stakeholders, yet this is what was truly needed
  - View of program success is static (we needed to revisit our view of success and what is needed beyond stated expectations)
  - When deliverables were submitted timely, the client is happy (we did not pay attention to the real goal beyond what was initially communicated . . . fluid process with true intimacy with the client)

Reality: Proper engorgement strategies are a key determinant for program success.

5.4.2 Tailoring Factors

PMI in its PMBOK Guide seventh edition puts emphasis on the need for tailoring project and program delivery approaches. This tailoring journey is a common sense one that begins with selecting the delivery approach, finding the right ways to customize or tailor that to the target organization or customer, then take that down to the program/project level, and ultimately go through the classic cycle of continual improvement. This is quite sensible and useful for dealing with the ongoing changes that surround the way of working and the delivery of future program work.

This tailoring is such a big priority in program work, so we need to take the time to do this properly to fit the needs of the stakeholders and the expectations of the program:

- Program delivery approach (hybrid future)
- Organization's culture and ways of working (have to be careful to adjust strategies around how people work best)

- Program/project nature, complexity, and stakeholders (including politics, agendas, and attitudes)
- Stakeholders' aptitude for using programs as an opportunity for learning (using programs and projects as labs is a great way to encourage learning and contribute to adding stories for future success)

Review Questions

Parentheses ( ) are used for Multiple Choice, when one answer is correct. Brackets [ ] are used for Multiple Answer, when many answers are correct.

_____ is a classic myth around program success as highlighted in the pharma case study.

( ) Ability to achieve outcomes
( ) Client exchanges were friendly
( ) Deep understanding of stakeholders expectations
( ) Sensing true needs of stakeholders' engaging

Which of the following points clearly illustrate the need for proper stakeholders' engagement? Choose all that apply

[ ] Stability of requirements in alignment with SOW
[ ] Key stakeholders change across the program life cycle
[ ] Program key components have demonstrated benefits
[ ] Witnessing that the deliverables are done yet the impact was not illustrated

Which is of the following is a contributor to customizing the strategic engagement of stakeholders?

( ) Repeatable patterns of delivering project work
( ) Organization's culture and ways of working
( ) Executive management's whim
( ) Using state-of-the-art technology

5.5 SENSING AND RESPONDING

Key Learnings

- PMI strategic journey story to highlight the centricity around stakeholders
- Learning organization's support for sensing and responding (put the knowledge focus and management hat on)
- Developing agility in program leadership (great future quality to build on)
- Balancing, exploiting, and exploring attributes in program leaders (short term and long term that includes experimentation)

5.5.1 PMI Strategic Journey Story

PMI is the source and resource of many of the standards and practices of portfolio, program, and project management. The perceptions reflected in figure 5.9 attempt to illustrate this journey. From building the profession in order to build the practice, to shifting and expanding to the wider folks who get involved in projects at large, and then to the recent years' strong focus on change makers.

	Early Days Professional	2017 Practitioner	2020 + Change Makers
PMI Strategic Journey	• Practitioners • Organizations • Academia	• PM Practitioners • Aspiring PMs	• Professionals • Youth • Students • Entrepreneurs

FIGURE 5.9 Profession Story

You could use this analogy to how to reflect on your own change journey and impact creation professional and personally. Looking at the dynamics of a journey such as PMI's, one would realize the need to be connected to the stakeholders around us to help us rethink existence impact, services, and how to effectively connect to stakeholders.

> **TIP**
>
> Leaders should remain connected to their stakeholders. Closely sensing their changing needs and remaining responsive ensures valuable positive impact.

5.5.2 The Learning Organization's Sensing and Responding

- Learning organizations create safe places to play (nothing is as important)
- Experimenting is everything (creative mode)
- Continuous production (deliver while excelling in the process to achieve the edge needed)
- Early stakeholders' input (so critical and can't be forgotten such as when you use MVPs in agile delivery)
- Humility and transparency drive the continuous learning culture (leaders and team members who know there is more to learn and are hungry to see themselves change and develop)
- Empathy drives customer alignment mindset (our goal is to provide expertise to help the clients in seeing what they won's see) Figure 5.10 is an example of empathetic and continual learning that is at the center of strengthening the future leaders' sensing and responding capabilities.

FIGURE 5.10 Safe Connecting
Source: sasint / 224 images.

5.5.3 Agility in Program Leadership

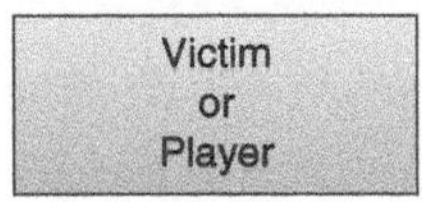

There are two positions leaders could take that have a vast difference in impact on their professional and personal success. These two ends could be the victim or the player positions. Organizational leadership has been shifting for a while. Program leaders need to be part of that shift. They got to shift to the player space where there is less blame, minimal finger pointing, or easy ways out.

Take the ownership that is central to one's adaptability (solutions mindset):

- Operating as a coach (player position)
- Stepping back from volunteering any solutions (pause for a moment)
- Open risk appetite (as the charter might indicate helps with escalation clarity)
- Collaboration qualities
- Emotional intelligence continual enriching (being put in tough situations and pushed to the edge of comfort zone)
- Operating with ownership mindset (ensure you remember the conductor and holistic leader responsibility)

5.5.4 Exploiting and Exploring Balancing Attributes

Figure 5.11 shows those exploiting and exploring attributes and what could be achieved when they are in balance. These are in continual tension in the work of the program and in the execution of organizational strategy. These also tie to the risk appetite that key program stakeholders have and that the program manager is exhibiting. Finding the right balance is an art and a science and requires sensible use of the head and the heart needed for effective program decisions.

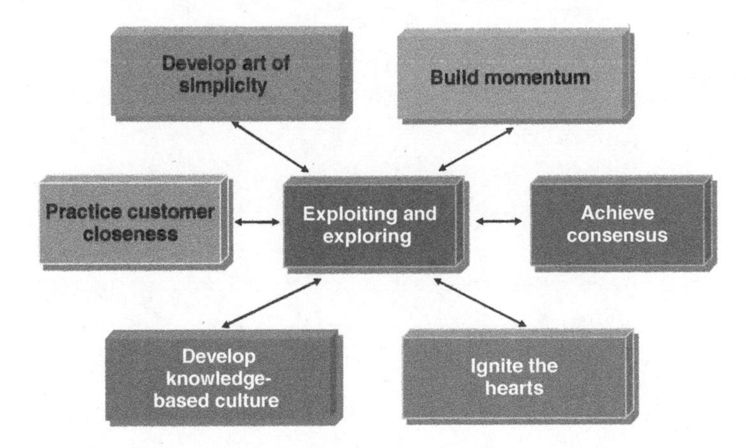

FIGURE 5.11 Exploiting and Exploring Attributes

Programs are complex enough. Thus, there is no need for added complexity. It is important for PMs to build coalition, drive consensus, develop the head–heart connection, and ensure that it is working, understand the need for and use knowledge, and ultimately achieve a balance between the short and long term with the customer in the center.

> **TIP**
>
> Program managers should develop the muscles of exploiting and exploring. They should also balance both in relation to the stakeholders and program needs.

Review Questions

Parentheses ( ) are used for Multiple Choice, when one answer is correct. Brackets [ ] are used for Multiple Answer, when many answers are correct.

_____ is an example of what contributes to enhanced organizational sensing and responding.

( ) Careful experimentation
( ) Humility and transparency drive the continuous learning culture
( ) Lessons learned at program closure
( ) Customer is always right slogan

Which of the following contributes to increasing the level of agility in leading program work? Choose all that apply

[ ] Stepping in with ideas and solutions
[ ] Risk appetite
[ ] Operating with ownership mindset
[ ] Using victim's language

Which is of the following are examples of balancing exploiting and exploring attributes?

( ) Increasing the program team's technical quotient
( ) Focus on immediate results
( ) Practice customer closeness
( ) Using complex roadmaps

CHAPTER 6

Power Skills

The future of work is changing. The recent years have shown us major scale disruptions in how we work, connect, and execute our work results. Although there have been immense explosions in technological innovations, it has never been clearer, more than ever, the criticality of soft, people, or social skills. PMI and other professionals around the world have been shifting to recognizing these skills as ***Power Skills***. There is no surprise in the choice of these two words, as it has become more evident that these skills are the ones with the true power behind successful strategy execution. They are the future difference maker.

This chapter addresses the context of why it is valuable to explore the principles of creating the mindset shifts necessary of mastering these skills. Understanding the important balance between technical and human skills is an important muscle to develop. Building a strong awareness of the attributes and value proposition of becoming a PMI certified Program Management Professional (PgMP), contributes to the leading by example role program managers could play.

Key Learnings

- Role of diverse skills in future of work has exploded (foundational point)
- What is skills revolution? Why should program manager's care? (Your call to action)
- Brené Brown and John Maxwell examples for different ways of leading and mindset shifts (Great leaders to help us stretch)
- Unlearning, skilling, upskilling, reskilling: what are differences and how challenging some of this could be? (Sponge like, continuous effort to challenge our assumptions)
- Build skills momentum (continuous improvement and adjustment)

6.1 THE SKILLS REVOLUTION

This power skills revolution is here. It is important to develop these skills and to have a roadmap for continuous skills improvement. It is also valuable to know that the skills revolution is coupled with associated dynamics of unlearning, skilling, upskilling, and reskilling. This enables the program managers to learn how to use those power skills to drive and impact their programs' future success.

Program Management: Going Beyond Project Management to Enable Value-Driven Change, First Edition. Al Zeitoun.

6.1.1 The Diverse Skills in the Future of Work

Figure 6.1 shows the future opportunity that we have in projects and programs. The wider set of skills shown will require continual investment in their development and practice. This future of work skills builds on the professional standards, readings, and practicing of the principles, while focusing on being helpful to the team, building trust innovating, sensing, and establishing stronger connections to stakeholders.

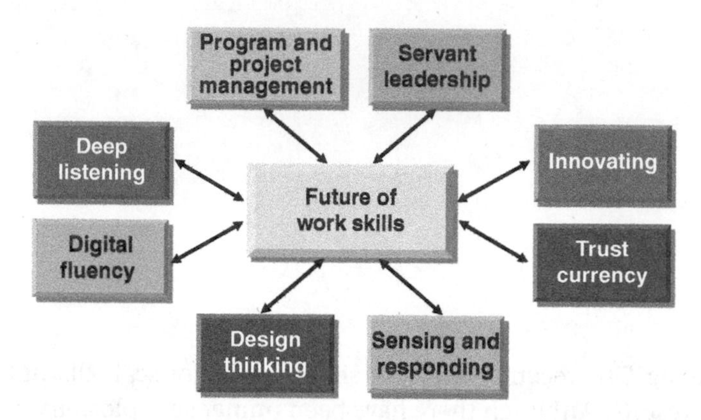

FIGURE 6.1 Future of Work Skills

Among these skills, which directly affect the effectiveness of problem solving, is the classic value of design thinking in finding the true needs of customers. Leaders in the future will find friendly utilization of digital and couple that with deep listening that is seen from a customer view point.

> **TIP**
>
> Leaders in the future will have to develop a mix of skills that enable creative problem solving with trust as a foundation to enable digital and human qualities

6.1.2 The Skills Revolution

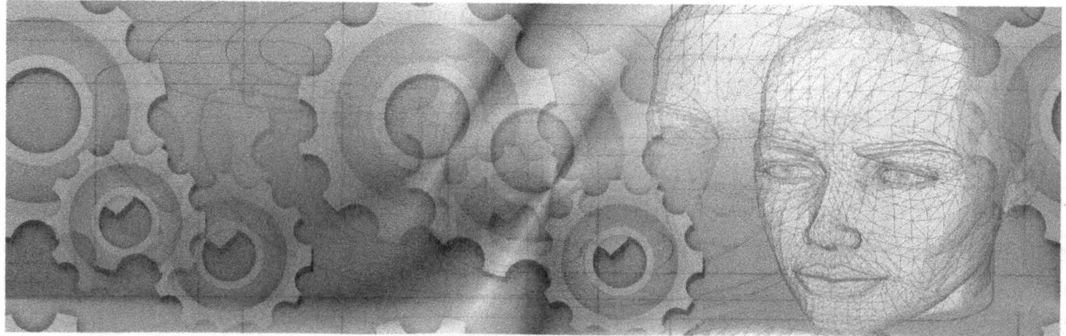

Source: kalhh / 4296 images.

The Revolution

- Cultural Implications (how we work, including the stretching in virtual and digital)
- Fluidity in Ways of Working (can't rely on any of the classic ways of working)

The Program Managers

- Speak executives' language (programs are strategic)
- See holistic end-to-end context (you provide integration and linkages)

6.1.2.1 Ways of Leading and Mindset Shifts

Brené Brown:

- There is no innovation and creativity without failure. Period.
- Through my research, I found that vulnerability is the glue that holds relationships together. It's the magic sauce (shifts our views and gets us comfortable to support creating aligned networks).

John Maxwell:

- A leader is one who knows the way, goes the way, and shows the way.
- Leaders must be close enough to relate to others, but far enough ahead to motivate them (great balance that we continuously should work on, relate, yet have a distance to see the vision and connect the dots to the program).

6.1.2.2 Unlearning, Skilling, Upskilling, Reskilling

Source: Dark_lone_nature / 61 images.

- The future is T-shaped skills rich (deep and general mix, to play different roles and wear different hats)
- Ongoing process unlearning and learning (get bias out of the way)
  - Growing library of changing skills
  - Integration, speed, and diversity matter most (appreciating different views and their contribution to program success)

6.1.2.3 The Power Skills Categories

In PMI's pulse of the profession research[1] data from nearly 3500 professionals across the globe was analyzed. The four top skills categories that contribute to project success were found to be ***communication***, ***problem-solving***, ***collaborative leadership***, and ***strategic thinking***. The research also addressed the areas of success these groups

[1]Based on PMOI's Pulse of the Profession 2023 Report – Power Skills, Refining Project Success.

of skills have improved. For organizations that prioritize power skills, they were able to meet their business goals better, face less scope creep, and minimize budget loss.

> **TIP**
>
> Leaders will develop T-shaped skills, unlearn and learn, and creatively build their communication, problem-solving, collaborative leadership, and strategic thinking.

6.1.3 Building Skills Momentum

Figure 6.2 shows the building blocks for creating the skills momentum. Program managers will need to operate effectively in the future of work.

The key is creating momentum that is anchored in a healthy culture, uses principles to help program managers nurture the coverage that they need, and uses dynamic skills library and scales diversity with trust and respect. This momentum also reminds us that we are all in a software business.

Culture: rewarding new behaviors, extreme collaboration

Principles: balance with processes, strategic role of programs

The human: dynamic skills library use, diversity at scale

The digital: end-to-end, all in a software business

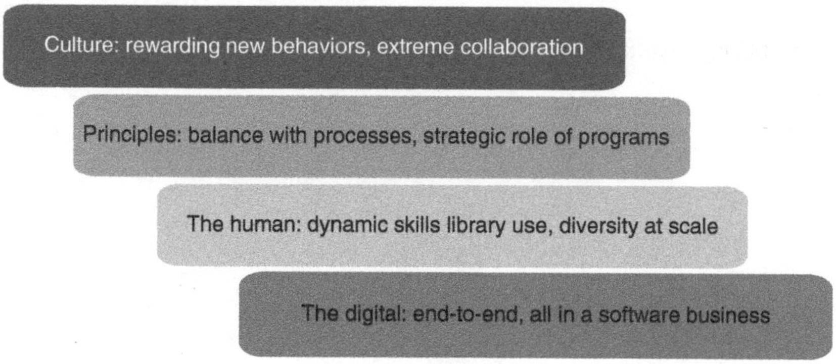

FIGURE 6.2 Skills Momentum

Review Questions

Parentheses ( ) are used for Multiple Choice, when one answer is correct. Brackets [ ] are used for Multiple Answer, when many answers are correct.

______ is a critical anchor to the skills revolution that program managers should care about and lead.

( ) Command and control culture.
( ) Fluidity in the way of working.
( ) Speak the team's language.
( ) Focus on the technical program details.

Brené Brown said "There is no innovation and creativity without failure." Why is that critical mindset shift for future programs success?

Choose all that apply

[ ] Provides organizations more programs control.
[ ] It enhances our risk appetite.
[ ] Creates the safety needed for better ideation.
[ ] Work becomes more strategic.

Which of the following building block of creating momentum will help breakdown the program silos?

( ) Higher focus on building deep technical skills.
( ) Focus on structured processes use.
( ) A culture that rewards new behaviors.
( ) Only accept consensus as the decision-making tool.

6.2 NOT SOFT SKILLS ANYMORE

Key Learnings

- Best scenario – program managers are at center of the skills revolution (what is your call to action?)
- Put together a list of key skills that should be prioritized across programs (use across phases, etc.)
- Develop resilience mindset
- Build collaboration routines (the intentionality and intensity of collaboration)
- Anchor Power Skills with Art of Grateful Leadership by Judy Umlas

6.2.1 Program Managers' Role

- Ownership mindset for instilling power/human skills (how can I instill that and have value in the center, reflected by making things happen highlighted in figure 6.3)
- Success stories of mapping skills to programs' success (this is what makes things stick with the team)
- Lead by example while observing talent in action (you are at the center stage and are being watched in action)
- Embrace diversity (one of your most key qualities, participants working on the program from across silos, entities, regions, etc.)

FIGURE 6.3 Ownership for Success
Source: geralt / 25607 images.

6.2.2 Key Programs Success Skills

As in Figure 6.4, the key program success skills can be grouped into 3 buckets of skills: unlimited potential of the team, polishing resilience around the dynamic nature of programs, and growing the natural skills intentionally, making sure that the few critical human qualities are remaining effective in the future, and will be developed well.

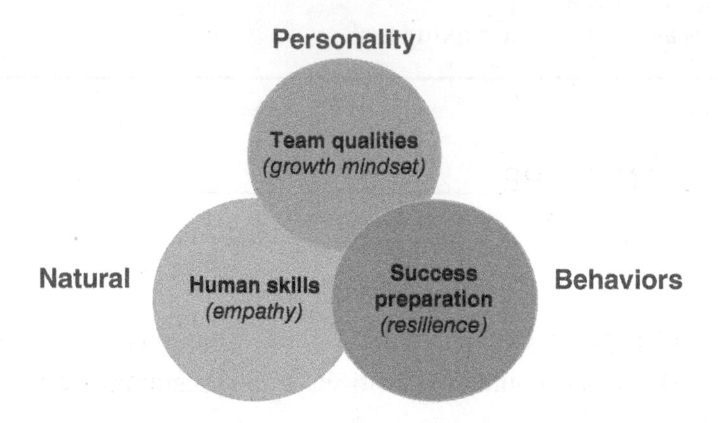

FIGURE 6.4 Success Skills

> **TIP**
>
> The future belongs to leaders who exhibit a growth mindset. These leaders will use their EQ to make dreams a reality

6.2.3 Resilience Mindset

Successful program managers realize that the higher the resilience, the lower the consequences of unexpected programs' risks.

Reminders:

- Vulnerability is everything in being ready to learn and adapt
- Bouncing back quickly and effectively (great quality to remain positive in front of negative consequences and jump on the opportunities)
- Reframing challenging while leading with trust and empathy
- Role of partnerships (cherish the learning gained and the associated agility in leading)
- Value of agility, creativity, and strategic focus (driving how we think and operate and illustrate our leadership)

6.2.4 Collaboration Routines

Figure 6.5 highlights that in the future of work, ***Collaboration = Innovation Success.***

Healthy Routines:

- Daily check-ins
- Team-led program responsibilities (cascading for higher commitment)
- Positive interactions' behaviors
- Simple rules: lightning rounds, five-minute rules, summarizing (e.g. Pat Lencioni and his published work in *Death by Meetings*)

FIGURE 6.5 Collaboration Routines
Source: geralt / 25607 images.

6.2.5 The Art of Grateful Leadership[2]

- The acknowledgement principles strengthen the impact of the program manager (simple and inexpensive tool for the PM)
  - Releases program team's energy (empowered)
  - Truthful and sincere ways of recognizing team performance sustain program benefits
  - Research shows positive emotional and physical value (no surprise as humans would thrive under specific acknowledgement/relatedness)
  - Tailoring acknowledgement for best results

> **TIP**
>
> Program managers are practitioners of the Power of Acknowledgment. They use this power skill to relate at a deeper level to their teams and achieve superb outcomes

Review Questions

Parentheses ( ) are used for Multiple Choice, when one answer is correct. Brackets [ ] are used for Multiple Answer, when many answers are correct.

Which of the following are examples of balancing exploiting and exploring attributes?

( ) Increasing the program team's technical quotient.
( ) Focus on immediate results.
( ) Practice Customer Closeness.
( ) Using complex roadmaps.

(continued)

[2]Adopted from the Seven Principles of the Power of Acknowledgment, by Judy Umlas.

(continued)

_____ is an example of a program team quality that would present a DNA in support of program success.

() Always ensure detailed requirements before starting
() Developing a growth mindset
() Ensuring the practice of customer is always right
() Escalate as much as possible to program sponsor

Which of the following contributes to building a resilience mindset for the program team? Choose all that apply

[ ] Showing a fearless face
[ ] Reframing challenging
[ ] Using partnerships to aid in managing unexpected risks
[ ] Strengthening tactical focus

Which of the following is related to the power of acknowledgement principles?

() Thank the team members for everything
() Program team works harder
() Positive emotional and physical value
() Use the same acknowledgement style for consistency

6.3 THE PROGRAM SUCCESS LINK

Key Learnings

- Reflect on Bill Gates' comment: "It's fine to celebrate success but it is more important to heed the lessons of failure." (Learning from what not to do could be very effective)
- Create alignment, reinforming the role of communication and culture (exemplify and live the skills)
- Hidden barriers to moving programs forward
- A look at EFQM model and skills that matter

6.3.1 Celebrating Success

There does not seem to be enough attention given to celebrating success. As illustrated by the joyful body language highlighted in Figure 6.6, program managers should pause and create opportunities throughout the program lifecycle to conduct such celebrations. We could also celebrate failure if learning has been achieved as it is likely to have a long-term positive strategic impact on the business growth. Timely lessons learned are very valuable, and balancing the learning from both success and failure is very important muscle to develop.

Why is learning from program failure valuable?

- Mind is wired to remember mistakes
- When safety is in place, team can openly learn
- Creating timely celebration directly motivates value-focus (use this to connect while team is still sweating and is in the trenches of the work being done)

FIGURE 6.6 Success
Source: JillWellington / 2139 images.

6.3.2 Role of Communication and Culture in Alignment

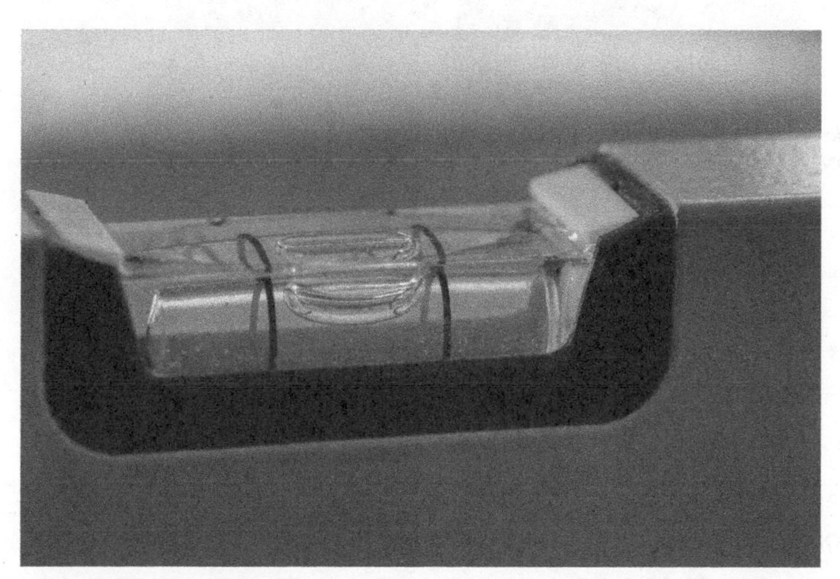

Source: planet_fox / 2756 images.

- Proper sponsorship empowers decision-making (role of the sponsor sets the foundation for success)
- Team admitting to making mistakes (humbleness that we don't have all eth answers)
- Sharing estimating data (example in estimating, create the buy-in)
- Team-centered leadership (team experience needs to stay intact)
- Program management viewed as a strategic competency

Barriers to Moving Programs Forward

- Bureaucratic centralization of authority in hands of a few (no go for the next decade)
- Insecurity at executive levels (may expose them)

- Fear of implementation failure risks (depends on how leaders emphasize learning from failure)
- Lack of tolerance for diverse views of virtual team members and partners (no go, we got to cherish and ask for complementing our views)
- Infrastructure exists to filter bad news from reaching the top (show the ROI and the loss that would happen if we don't share this news as early as possible)

6.3.3 EFQM Model and the Skills that Matter

Perceptions affect the evaluation of success and the ability of the program manger to lead in this future of work. Program managers should manage and drive the stakeholders' perceptions in evaluating success in order to see the programs' true benefits. Changing perceptions is empowered by dynamic reskilling and upskilling. A blended culture builds on co-creation ideas and the more we integrate across cultures, the higher we will have the skills needed for program success.

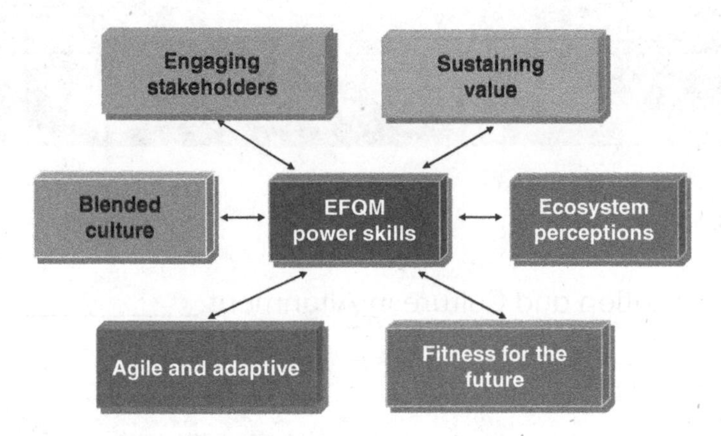

FIGURE 6.7 Power Skills in EFQM.
Source: Adapted from EFQM.

> **TIP**
>
> Power skills provide us with the fitness for the future. Sustaining program value is built on a culture that enables ecosystem adaptability and effective engaging.

Review Questions

Parentheses ( ) are used for Multiple Choice, when one answer is correct. Brackets [ ] are used for Multiple Answer, when many answers are correct.

_____ illustrates the role of culture and communications in creating strong alignment.

( ) Documenting all program interactions.
( ) Supporting the program team with simple roadmaps and charters.
( ) Continuously insisting on holding program meetings.
( ) Leader-centered leadership.

Which hidden program barriers continue to affect programs' success?

Choose all that apply

[ ] Alignment across business units.
[ ] Fear of risks of implementation failure.
[ ] Sharing bad news.
[ ] Lack of tolerance for diversity.

Which of the following EFQM skills is mostly enhanced by the program manager's assumption management abilities?

( ) Fitness for the future.
( ) Blended culture.
( ) Ecosystem perceptions.
( ) Sustaining value.

6.4 POWER SKILLS MASTERY

Key Learnings

- Use Pat Lencioni's 5 Dysfunctions of Teams to build the foundation for program team's high performance (ensuring that the team is focused)
- Levers that program managers can use to shape culture and climate for skills mastery
- Reflect on Capability Maturity Model Integrated (CMMI) and other maturity models (helpful to support mastery and excellence)
- Self-leadership and resilience (including managing stress)

6.4.1 The 5 Dysfunctions of Teams

Figure 6.8 shows Pat Lencioni's well-articulated model for what could lead to dysfunction in program teams and how to overcome that is a must-have tool in the arsenal program managers need. Going up the cohesive team pyramid, we reach the focus on the joint results, which is the integration we want to see across program teams. On the other hand, in the dysfunctional pyramid, we can see the negative behaviors that create the politics, games, and then no attention is left for the results that matter.

The most important layer of this pyramid could be the second one, around conflicts and their use to protect trust. Teams that are unhealthy are unable to deal with conflicts well and timely. This journey is not a linear exercise as you could head up and then down this Lencioni pyramid.

TIP
Leaders should build their teams on a foundation of trust that allows critical discussions to take place. This safety commits the team to owning results.

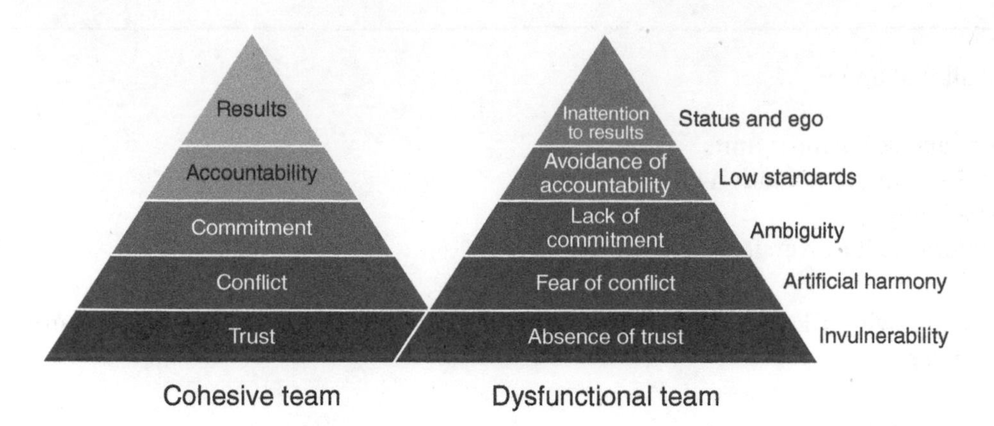

FIGURE 6.8 Impact of Program Teams Dysfunction
Source: Adopted from the Five Dysfunctions of Teams, by Patrick Lencioni.

6.4.2 Levers for Skills Mastery

Source: WOKANDAPIX / 1264 images.

Reminders for Mastery

- Leave the most impactful leader's shadow (how you come across, walking the talk, practicing the skillsets)
- Continual learning appetite (getting that across to the team)
- Be yourself more with power skills (while remaining authentic as a leader)
- Take on more complex challenges (this is how you can stretch, putting oneself in tough complex situations)
- Practice dynamic movement from individual to a whole system view (e.g. being on the balcony versus the dance floor learned in previous lessons)

6.4.3 Maturity Models

- Mastery is a journey of repeatable successful patterns again and again (not a one-time thing)
- All maturity models including CMMI, Kerzner's, and others highlight need for continuous improvement practices
- A supportive culture for program excellence is a must (operating from an owner mindset to set the environment for excellence)

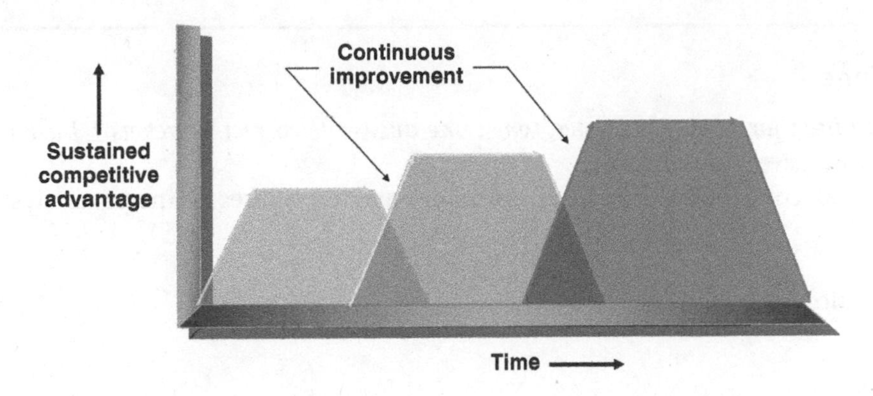

FIGURE 6.9 Maturity Journey for Mastery

6.4.4 Self-Leadership and Resilience

Figure 6.10 reminds us that one of the contributors the top power skills of the future is stress management. The more programs complexity increases and expectations grow, the more valuable this skill will be. Stress management consists of the following competencies:

- **Flexibility:** adapting to change effectively
- **Stress tolerance:** successfully coping with stressful situations
- **Optimism:** having a positive outlook

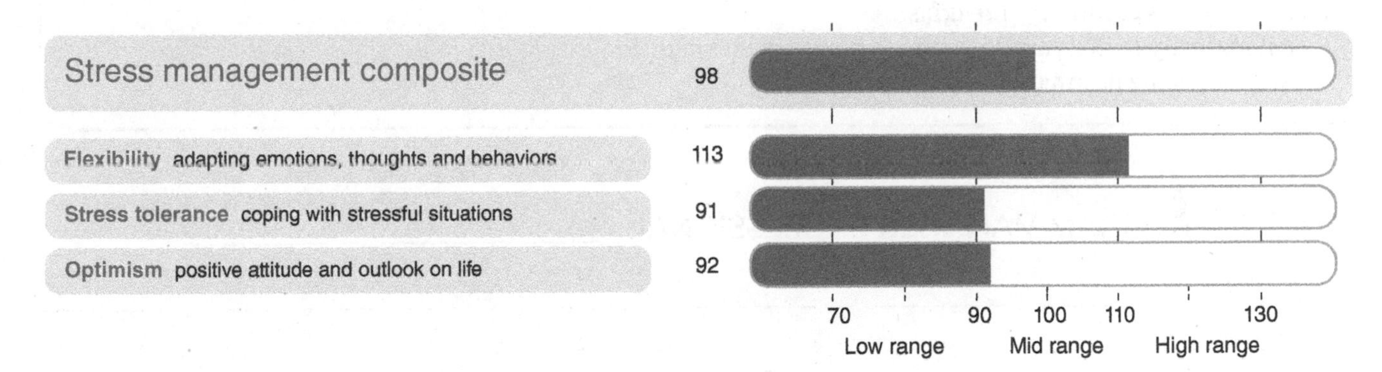

FIGURE 6.10 Sample Stress Assessment
Source: Adopted from EQ-i$^{2.0}$ in measuring five distinct aspects of emotional and social functioning.

> **TIP**
>
> Future leaders will invest in their stress management. This will give them the ability to adapt and use their positive attitude to drive change.

Review Questions

Parentheses ( ) are used for Multiple Choice, when one answer is correct. Brackets [ ] are used for Multiple Answer, when many answers are correct.

_____ is what could contribute to artificial harmony and thus directly increases dysfunction on the program team.

( ) Attention to results.
( ) Fear of conflict.
( ) Commitment.
( ) Lack of trust.

Which of the following directly contributes to increasing your program human skills mastery? Choose all that apply

[ ] Stepping in others' shadows.
[ ] Learning appetite.
[ ] Ability to zoom in and out.
[ ] Electing to lead simple programs.

Which of the following is an example of the maturity modals' contribution to skills mastery?

( ) Level 5 guarantees programs success.
( ) Strategy eats culture for breakfast.
( ) All maturity models are continuous improvement centric.
( ) Mastery is the destination.

6.5 THE PROGRAM MANAGER PROFESSIONAL

Key Learnings

- Program manager as organization's benefits strategist (biggest shift in the program manager role)
- Why does professionalism in program management matter?
- Understand practices behind becoming certified (how to get there?)
- How to overcome illusion of certification means final accomplishment? (Be more excited about the process)
- Identify practical tips for creating program managers' mindset needed to become certified (speaking the language of the profession such as everything this certificate has been focused on)

6.5.1 Benefits Strategist

Source: Positive_Images / 151 images.

As in the nice pic of the masterful chess player, this gives us the connection that we can develop the strategic qualities of a Benefit Strategist across generations and levels of expertise.

- Critical future program manager role
- Ownership of benefits lifecycle story (you are the connector we rely on that the right stakeholders are connected)
- Thinking across hats ability (thinking for a change again)
- Diversity of views
- Professionally responsible in the practice of the program work (what is proper and what is not, etc.)

6.5.2 Value of PMI-PgMP

- Professionalism matters (confirms the program manager's commitment)
- Consistent use of program language (commonality helps the teams)
- Maturity of practices across ecosystem (many players in the mix)
- Creates credibility and continence across team and teams of teams
- Successful handling of uncertainty and complexity (the outcome of being well trained, practiced, and coupling that with completing the certification)

6.5.2.1 Becoming Certified

As highlighted in Figure 6.11, pursuing such an important program management professional PgMP certification requires rigor, preparation, and integrating multiple areas of learning and expertise together to demonstrate the leader's abilities.

FIGURE 6.11 Pursuing Certification
Source: knowledgetrain / 3 images.

- Operating as a program manager
- Building program mindset (as in our journey in this program)
- Training and coaching opportunities
- Comprehending of PMI's Program Management standard (capturing gaps in your understanding and prep)
- Completing a rigorous certification approval process
- Passing of panel review and certification exam

6.5.2.2 Certification Is Not a Final Accomplishment

Source: piviso / 63 images.

- Myth behind certification being an ultimate goal (way of working requires the ongoing learning process)
- Role of consistent behaviors
- Evidence in results
- Impact on program team (your approach has to change and you should show how you are taking obstacles out of the way)
- Culture building contribution

6.5.3 Practical Tips for the Program Managers' Mindset

Figure 6.12 summarizes the components necessary for the program manager to build the possible mindset for success. The program manager has a mindset that is able to: be ***strategic***, build the role of the sponsor and that partnership, express ***ownership*** including in building the team, ***engaging*** stakeholders, invest in ***coaching***, and come across as the holistic leader needed as a program manager and a certified professional.

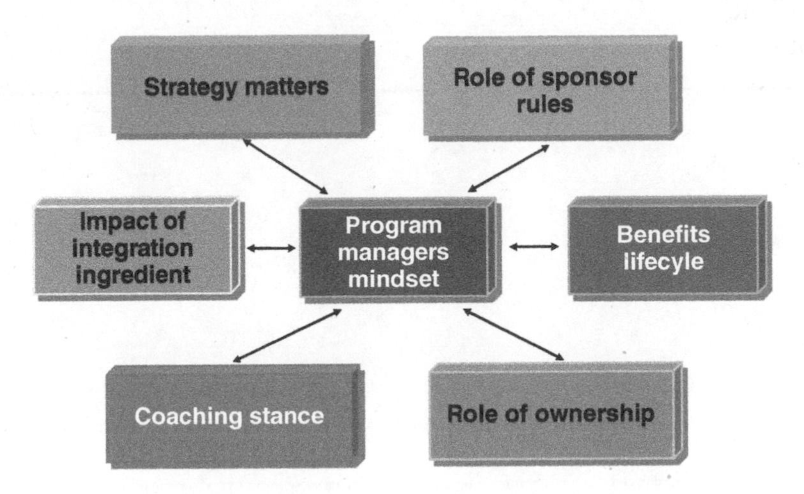

FIGURE 6.12 Mindset Shifts

> **TIP**
>
> To become a PgMP, the leader should understand strategy, have clear abilities for owning benefits, work well with sponsors, and coach effectively.

Review Questions

Parentheses ( ) are used for Multiple Choice, when one answer is correct. Brackets [ ] are used for Multiple Answer, when many answers are correct.

_____ is an example of building the program manager's ability of being the program's benefit's strategist.

( ) Having strong views.
( ) Thinking across hats ability.
( ) Focusing on the planning phase.
( ) Building a bias for action.

(continued)

(continued)

Which of the following are key steps for becoming a PMI Program Management Professional? Choose all that apply

[ ] Getting good program reviews.
[ ] Building the program mindset.
[ ] Comprehending of PMI's program management standard.
[ ] Being likeable.

Which of the following is a critical attribute in the program manager's mindset?

( ) Increasing the reacting capacity.
( ) Focus on deliverables.
( ) Integration ingredient.
( ) Directing the team.

CHAPTER 7

Digitized Future

The digitized future is here. With ChatGPT as an example of advancements in Artificial Intelligence (AI), which is going to directly impact the work of program teams, their access to meaningful information and trends, and the efficiencies program managers will have at their disposal. Program managers must be focused on enabling the vital importance of the vast enablers in our digitized future. They should understand the impact that AI will continue to have on the future of programs.

As a program manager, you have an opportunity in this chapter to get immersed into the importance of digital solutions in better understanding of accurate programs' context. You will learn the direction that digital disruption is taking and what the implications are expected on programs and the role of the program manager.

You will also learn the Human 2.0 equation and its use in creating the connected and human future of work. This will be valuable in developing the digital balance across program teams and the increasingly growing hybrid model of working.

Key Learnings

- Is AI going away anytime soon? Very doubtful
- How vast is the mix of digital solutions available to program manages? (How does that affect your role as PM?)
- Every organization is a software organization (This has been evident for years and now is becoming mainstream)
- Digital integration and the difference making (This digital universe directly impacts PMs coupled with the previous power skills we discussed)

7.1 AI IS HERE TO STAY

The future of work is highly digitized. AI could become a difference maker. For years, the technology has been perfected and has now reached a stage where the dots are being fully connected and thus valuable insights are captured, shared, and very fast. This is a very special program/project work enabler that is for sure here to stay.

Program Management: Going Beyond Project Management to Enable Value-Driven Change, First Edition. Al Zeitoun.

7.1.1 AI and Programs

AI can be used in managing programs to increase efficiency of operating intelligently, accuracy of outcomes, and to enhance decision-making.

Why is AI affecting programs and not going away?

As illustrated in figure 7.1 regarding the value of AI, there is potential for massive saving of time wasted in areas that don't directly tie to achieving programs success.

- Exploding across industries
- Massive data could be analyzed in seconds for refined programs' decisions (speed and quality of these decisions)
- Large AI investments transforming society's way of living and working (affects our quality of life)

Multiple program work areas could benefit from AI. Enterprise Risk Management (ERM), as previously addressed, is one of those critical program areas. Machine learning and its algorithms could enhance the program manager's ability to forecast the probability and impact of program risks and even suggest the most effective handling strategies. The powerful predictive analytics could also help the program manger to plan better. In addition, the automation of reporting will enhance its quality, identify patterns fast, and give back time for the program manager to be more strategic and become able to invest with confidence in decisions that matter.

> **TIP**
>
> Leaders of future programs will operate differently. AI is a difference maker and will enhance the quality of where leaders spend their time and the outcomes created.

FIGURE 7.1 AI is Here to Stay
Source: geralt / 25607 images.

7.1.2 Digital Solutions for the Program Manager

Source: geralt / 25607 images.

- Numerous planning and control software
- Automated dashboards built on mature data analytics (ensuring seamless decision-making across teams)
- Mobile solutions for timely reporting and decisions (immediate and visual changes)
- Seamless program team interactions across globe (boundaries are falling out)

7.1.3 We are Software Organizations

The future is here. Across industries, we are becoming more digital, including Software as a Service (SaaS). Every business operates as a software organization. This is becoming a key attribute of organizations' DNA and has revolutionized everything we do and how we do it.

Software has affected everything we do as program leaders too:

- How we plan
- How we produce
- How we interact with customers and users (seamless)
- How we sell (simulate the experience)
- How we partner
- How we track our performance (fast ways to get back on track)
- How we learn and improve (learning organizations)

7.1.4 Digital Integration as Difference Making

Individual elements are great, yet integration scales outcomes and benefits fast and more effectively.
This enables us to:

- See customers in a different light
- Capture vast amount of useful data that digital allows us to understand and put to good use (insights and simplicity)
- Move to cloud, use of AI, and IOT, when linked to strategy, makes integration central to the digital future (capitalization on the integration, linking to business strategy, relating differently to our customers)
- For example, using C3.ai platforms
  - organizations able to connect data sets to drive useful insights
  - demonstrate programs' ***value***
  - achieve effective business decisions (faster feedback and assurance of the caliber of decisions)

Review Questions

Parentheses ( ) are used for Multiple Choice, when one answer is correct. Brackets [ ] are used for Multiple Answer, when many answers are correct.

_____ illustrates that AI is here to stay and directly contributes to strengthening programs' decision-making.

( ) The increasing need for openness.
( ) Massive data that need to be analyzed in seconds for refined programs' decisions.
( ) Government is expecting it.
( ) Team members love gadgets.

Which of the following are indicators that most organizations have become software organizations at their core?
Choose all that apply

[ ] Every organization likes the Amazon model.
[ ] The ways we track program performance.
[ ] Many program constraints.
[ ] How we sell and interact with customers.

Which of the following is a strong value contribution of enhanced digital integration?

( ) Humans are less needed in the future.
( ) Management demands it.
( ) Sense making out of massive amounts of program data.
( ) Creating a better perception of the organization.

7.2 THE DIGITAL EDGE

Key Learnings

- Use Simon Sinek's Infinite Game concepts to extract digital edge (reach the potential of digital)
- Build digital capabilities will be demonstrated using a Siemens simulation story (co-creating)
- Develop digital enterprise with low-code industry-specific applications (forecasting how that affects the way of working)
- Review enterprise risk management and its correlation to success in digital solutions (turning uncertainties into opportunities exactly like digital enables us to do)

7.2.1 Infinite Game

Mindset principles support achieving programs' strategic value.

Clarity of purpose, trust and its link to collaborative environment, survival, being courageous, and creating the safety for the potential of growth are all mindset qualities that lead to infinite growth potential for leaders and businesses. Among all the dimensions shown in Figure 7.2, the courage to lead while trusting the team is a fundamental value for the program leaders' ability to drive change and connect stakeholders. These ingredients enhance the limitless growth possibilities for the program team.

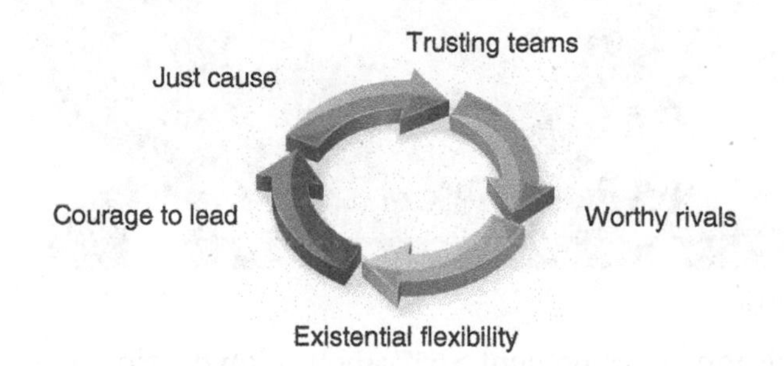

FIGURE 7.2 Mindset Principles
Source: Figure adapted from: http://www.tradeanatomy.com/content/2016/4/10/weekend-review-developing-a-routine.

TIP

Program managers are well quipped to play the infinite game. The nature of program work requires meaningful causes and the courage to lead the change

7.2.2 Digital Capabilities[1]

- Digital twin's virtual entity that mirrors physical world, turning data into insights that enhance quality of program decisions (great saving and allows PM to move faster)
- Digital threads connect digitally enabled processes, an integrated ecosystem of digital (integration across the multitude of capabilities)

[1]Adapted from Zeitoun, Al. Siemens Digital Industries Software white paper, Simcenter: the heartbeat of the digital twin, adopting a digital mindset to deliver and scale future innovative solutions.

- Digital threads are about strategic focus, speed, agility, and rigor
- Simulation supports efficiently in finding assets that can be reused, speeding up programs' innovation (commonalities in solutions and products to build on and learn from)
- Having a digital charter turns complex programs into much more manageable component projects (clarity of success ingredients)

7.2.3 Low-Code Applications

Source: geralt / 25607 images.

- By 2024, low-code application development >65% application development[2]
- Why is this a game changer for program managers?
  - Development autonomy
  - Ecosystem transformation (eliminating bottlenecks)
  - Closer idea to reality connection (core to the gap that would typically exist in strategy execution)

7.2.4 ERM and Digital Solutions Success

ERM, as reflected in figure 7.3, helps with developing an eye for value, using digital to help with clear dashboards to support transformation decisions, and clarity of direction. This makes it easier to experiment in an open environment where transparency prevails, with a sponge like mindset for learning and adapting.

> **TIP**
>
> Strengthening ERM capability with digital applications allows for enhanced decisions. The transparency created, accelerates change, and regains value focus

[2]Gartner estimate.

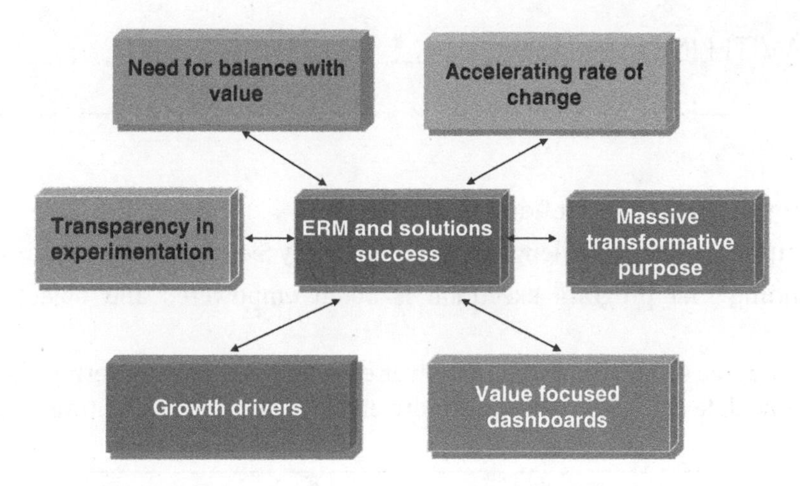

FIGURE 7.3 ERM and Digital Solutions Success
Source: Includes ideas adapted from Exo works.

Review Questions

Parentheses ( ) are used for Multiple Choice, when one answer is correct. Brackets [ ] are used for Multiple Answer, when many answers are correct.

_____ illustrates Sinek's infinite mindset principle that closely supports the changing role of the program manager.

( ) Worth Rivals.
( ) Courage to lead.
( ) Flexibility.
( ) Trust.

How do digital threads help us achieve speed, agility, and rigor?
Choose all that apply

[ ] By enforcing more structure across the program life cycle.
[ ] By connecting digitally enabled processes across an ecosystem.
[ ] By turning our program teams into tech savvy folks.
[ ] By creating the efficiently in finding assets that can be reused.

Which of the following is an example of how an ERM mindset enhances programs success?

( ) Increasing the controls around change.
( ) Focus mainly on applying governance rules.
( ) Supporting transparency in experimentation.
( ) Using purely business dashboards.

7.3 MANAGING WITH INTELLIGENCE

Key Learnings

- Build Trust Currency to strengthen confidence in Digital
- Examples to IOT implementation challenges as highlighted by Siemens Advanta
- Develop understanding that program execution is about empowered and objective decision-making (managing bias)
- Explore biases and the role of healthy trends to power enhanced program governance (manage with intelligence is built on how data can help us become more effective as humans and thus have more time for the relationships)

7.3.1 Trust Currency (TC)

- As attempted by figure 7.4, the more digital we become, the more valuable this future trust currency becomes (digital brings concerns such as cyber security, other team behaviors)
- Future of work is trust-central
- Investing in and protecting TC with critical conversations (safety built in the organization)
- Business success will require Humans who operate with **transparency** and who are not stuck (digital fluency supports this)

"Trust is the currency around which business foundations and the fabric of business occurs," Ed Bastian, CEO of Delta Air Lines.

TIP

Leaders should invest in building the TC. Building this muscle allows us to focus on what matters in the use of data for effective decision-making.

FIGURE 7.4 Trust Currency
Source: geralt / 25607 images.

7.3.2 Internet of Things (IOT) Implementation Challenges[3]

Source: Tumisu / 1245 images.

- Executives surveyed:
  - 90% can't determine ROI (a problematic data point)
  - 46% worry @ change mgmt.
  - 36% concerned @ cyber security
  - 55% fear modernizing operations (stuck and get comfortable in the old ways)
- Solutions:
  - Apply risk-driven strategy coupled with clear goals, agile change mgmt., digital adoption, and gradual asset introduction (key to successfully cascading strategy)

7.3.3 Empowered and Objective Decision-Making

A Booz Allen Hamilton secret sauce example is a strong illustration of how empowerment and effective decision-making go hand in hand:

- Set strategy guardrails (parameters only)
- Decisions cascading magic: executives share why and maybe some of what, then how is really engendering and empowering talent to help tell business leaders how to get there (conductor role helps with this)

The critical role of the program sponsor is to create right safe environment for reaching objective decisions (otherwise people will resist and this would create more limitations).

7.3.4 Healthy Trends for Enhanced Program Governance

Figure 7.5 highlights a set of healthy governance trends that contribute to a successful program environment. These healthy governance practices are seen at all levels, starting with different ways of working (WoW), light processes are the future, PMO become strategy offices will help with the end-to-end success of role of PMs, committing to the failure that helps us adapt better, getting the decision-makers to use technology in order to shift where they are spending their time (e.g. around strategic risks).

[3]IOT Challenges according to HBR Pulse Survey, 2019 and adapted from Siemens Advanta.

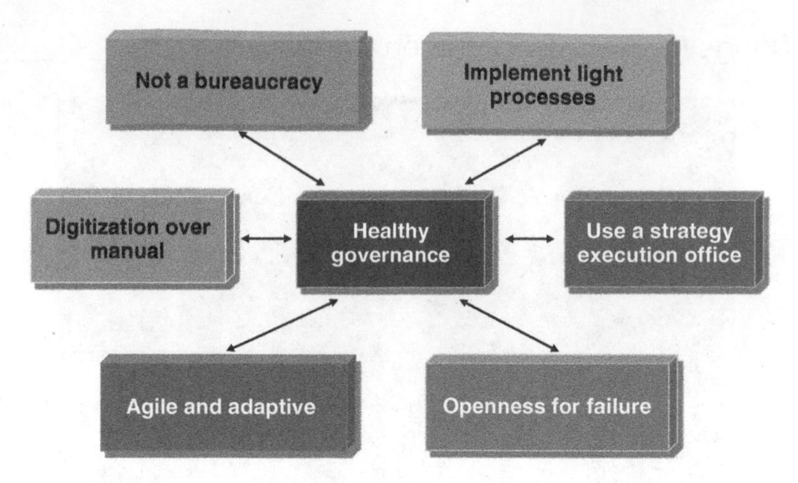

FIGURE 7.5 Healthy Governance Trends

Review Questions

Parentheses ( ) are used for Multiple Choice, when one answer is correct. Brackets [ ] are used for Multiple Answer, when many answers are correct.

_____ illustrates that the more digital we become, the more valuable trust currency (TC) is.

( ) Less conflicts among program team members.
( ) Investing in and protecting TC with critical conversations.
( ) Focusing on creating harmony.
( ) Cyber security breach.

Which of the following are useful solutions to deal with executives' concerns about the ROI and change management challenges of IOT implementation?

Choose all that apply

[ ] Agile change management.
[ ] Keep assets the same.
[ ] Plan a risk-driven strategy.
[ ] Leave the goals setting to the implementation team.

Which of the following could best own the excellence in governance practices while managing objectivity in decision-making?

( ) Steering group to increase use of the controlling measures.
( ) Strategy Realization Office.
( ) A team to drive simple manual processes.
( ) All stakeholders.

7.4 COMMUNICATING IS HUMAN

Key Learnings

- Reflect on Richard Feynman's statement: "I would rather have questions that can't be answered than answers that can't be questioned." (This was tackled from so many angles across the chapters and learnings, key message is: conflict and critical thinking are very important)
- Use powerful questions to ensure right degree of digitization is implemented
- Change the ways of working across program teams and across hybrid models of working (evident in how we collaborate)

7.4.1 Value of Questions

The world should not expect you as a PM to know all the answers:

- Value of a beginner's mind for future program managers (open ended, experimentation, curiosity is critical to the value of digital)
- Importance of questioning for critical thinking (are there limitations and why?)
- Asking powerful questions (come up with your own list of impactful questions, customize questions based on eth audience)
  - Is there another way of looking at this?
  - What would happen if . . .?
  - Is it always the case? When is it not?

As a future leader, you should show the ROI behind using questions to ensure achieving value.

7.4.2 Powerful Questions and Digitization

FIGURE 7.6 Powerful Questions
Source: Tumisu / 1245 images.

- As demonstrated by Figure 7.6, program managers should reflect and use powerful questions to ensure the achievement of the most valuable outcomes of digitization: Are the current operating models and processes supportive? (e.g. change management, learning, digital fluency, maturity)
- What is digital fluency of program stakeholders?
- What are safeguards for minimizing bias and achieving an effective human role? (How to maintain the valuable outcomes of digital achieved?)
- What are expected enhancements in program communication? (The key idea behind digital in communicating more effectively while we achieve the right decisions timely)

7.4.3 Hybrid Models of Working

As highlighted in the details of Figure 7.7, the traditional way of working is going to disappear in the future organization, shifting to hybrid, seeing co-creation as an ingredient to spread across the enterprise, while serving leaders use digital solutions to expedite their support to the teams.

Traditional program management	Hybrid program management
Program requirements are usually well-defined from start.	Program requirements are fluid and shaped based on value co-creation with business leaders and clients throughout the program.
Emphasis on benefits with tight governance practices + a directing program sponsor.	Servant leaders equipped with digital solutions enable expedited and effective decision-making.

FIGURE 7.7 Hybrid Working

Review Questions

Parentheses ( ) are used for Multiple Choice, when one answer is correct. Brackets [ ] are used for Multiple Answer, when many answers are correct.

_____ illustrates the importance of the beginner's mind in tackling complex program challenges.

( ) Higher focus on tactical program work.
( ) Asking powerful questions.
( ) Ease of digital use.
( ) Reaching consensus fast.

Which of the following is critical for the use of digitization readiness in program teams? Choose all that apply

[ ] Ensuring stronger governance.
[ ] Digital fluency of key stakeholders.
[ ] Number of global regions in the program.
[ ] Safeguards for minimizing bias.

Which of the following is a strong sign for the changing ways of working towards hybrid programs?

( ) Well defined requirements.
( ) Value co-creation with clients throughout the program.
( ) High involvement of program sponsors.
( ) Clearly established business objectives.

7.5 ACHIEVING BALANCE

Key Learnings

- Use the Organizational Health Index (OHI) to build and stain performance (foundation for what matters, achieving balance)
- Reflect on the Disney Imagineering example to demonstrate the balance between people and technology (completely sensitive)
- Human 2.0 equation will be used to draw in examples of creating the right people elements balance with this digitally rich program style (depending on where people are in their maturity journey)
- Use stories to complete the stop, start, and continue learning worksheets

7.5.1 Sustaining Performance

Source: 489327 / 3 images.

- As in figure 7.8, McKinsey looks at leadership, ways of working (WoW), and innovation (excellence depends on consistency)
- Program environment excellence depends on consistency in balancing, performing, and delivering with repeatable effective outcomes (critical to excellence)

The ladder that program managers could use as a reminder of balancing the creation of value with consistency of that achievement to create a path for organizational maturity.

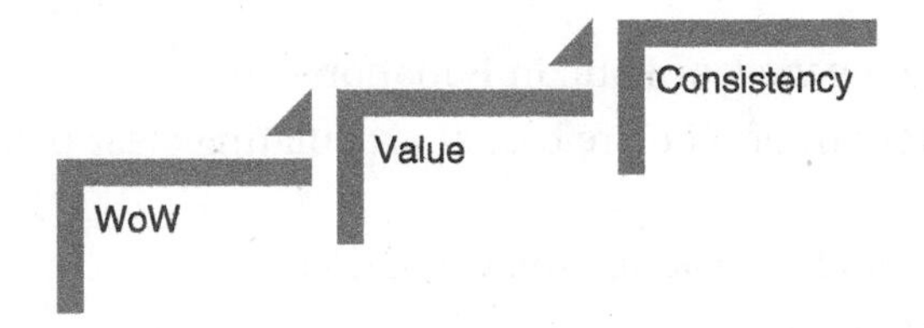

FIGURE 7.8 Sustaining Performance
Source: Adapted from McKinsey OHI

7.5.2 Disney Imagineering

Reflecting back on this story from a previous chapter, as in figure 7.9, this is always a fascinating story that inspires.

- Use of digital in closely integrating view of customers confirms unlimited potential of inspiring dreams (creating a flawless experience)
- Thread between people and technology creates need for new human equation for programs success (like living in the virtual space and the true utilization of the digital twin concept)

This example sets up the need to balance digital with human next.

FIGURE 7.9 Inspiring Stories
Source: dweedon1 / 15 images.

7.5.3 Human 2.0 Equation

The future is also ***human***, not just digital.

- Adaptive mindset requires new view of Human Equation
- Equation is centered on strength of who are best at articulating clear picture (insights, themes, views) . . . system minded as a PM
- 4 Equation elements surfaced clearly across the validation:
  - S = ***Sensing*** (very important)
  - I = ***Intentionality*** (experimentation and collaboration)
  - T = ***Thoughtfulness*** (having the time which is a direct plus of digital)
  - E = ***Energy*** (minimize impact of politics)

> **TIP**
>
> Future leaders focus on what matters most in the successful use of digital (the human side of the equation).

7.5.4 Stop, Start, and Continue Learning

A simple tool that is critical in bringing the beginner's mindset to the surface in the work of the program team.

- Need for a sponge mindset (squeeze out what you don't need, e.g. unlearning, relearning, etc.)
- Continual learning and programs' behavioral awareness and changes (intelligent ways of working and Guided Continuous Improvement (GCI), as reflected in figure 7.11, practice-practice-practice)

FIGURE 7.10 The Science of Program Learning
Source: geralt / 25607 images.

Stop	Start	Continue Learning

FIGURE 7.11 Learning Practice

> **TIP**
>
> Program managers should practice this simple science of program learning. They should have the courage to stop, start, and then continue to stay humble and learn

Review Questions

Parentheses ( ) are used for Multiple Choice, when one answer is correct. Brackets [ ] are used for Multiple Answer, when many answers are correct.

_____ illustrates the three building blocks of sustained program performance.

( ) Consistency, repeatability, and control measures.
( ) WoW, value focus, and consistency.
( ) start, stop, and continue.
( ) Value, strong alignment, and communications.

Which of the following are key ingredients of the Human 2.0 equation?
Choose all that apply

[ ] Toughness in program dialogues.
[ ] Sensing the changing needs of program stakeholders.
[ ] Ideating without limits.
[ ] Maintaining program teams' energy and passion.

Which is the value of the practice of stop, start, and continue?

( ) Strengthening the program team's structure.
( ) Focus mainly on catching failures.
( ) Creating a balance of focus on learning.
( ) Disregarding time pressures.

SECTION III

THE PROGRAM MANAGEMENT OFFICE (PMO) – THE STRATEGY EXECUTION ARM

SECTION OVERVIEW

This section addresses ways to stretch your program management and leadership capabilities further, while you uncover the critical principles of the Program Management Offices (PMOs) of the future. Your journey is heading towards practicing the very important future of work focus on creating value, cocreating, elevating program governance, and creating enterprise value by being the learning engine that drives the effective business cultures of the future.

You will get the opportunity to learn and understand how to create a value-driven mindset, the secret sauce of successful PMOs. You will learn a multitude of effective hybrid approaches to organizing your program's work, while governing with strategic agility and a balanced use of risk management practices. There will be examples and case studies that get you to deeply understand the importance and mechanics of getting the PMO to play a key role in operating as an End to End (E2E) enabler of capabilities, tools, and processes. A culture of learning will be demonstrated to help you consistently focus on program benefits and build the continuous improvement muscles.

SECTION LEARNINGS

- The importance of the PMO as the organization's key strategy execution arm
- How to take the time to step back and mark your calendar to think about the execution excellence practices and ideas
- Fitting the strategy execution and program delivery approach to what the program work needs
- Understand how to create outcomes and achieve business values that matter for your own business or for tomorrow's organizations
- Use the PMO mission and practices to build the impactful future learning organizations

Program Management: Going Beyond Project Management to Enable Value-Driven Change, First Edition. Al Zeitoun.

KEY WORDS

- Learning Culture
- PMO
- Hybrid Working
- Risk-based Governance
- Co-creating
- Value Mindset

CHAPTER 8

Value-Driven Programs and Hybrid Work

This chapter focuses on understanding the importance of designing programs as value-driven strategy execution vehicles. The leading entity of that design remains to be a Program Management Office (PMO) format. The name of such an entity could vary, yet a central entity for owning the continual cultural focus on value is critical to have. Part of the focus of this chapter is the importance of being sensitive to the delivery approach that is most suitable to a given program and organizing around that.

Program managers are at the center of this accountability for ensuring that our programs are value-driven. They are tomorrow's impact drivers.

Key Learnings

- Understand how to create a value-driven way of working
- Get introduced to the attributes of developing the value mindset
- Address the creation of trust in the PMO as the partner in driving strategic efforts
- Explore the secret sauce of building benefits management in the actions of the PMO
- Learn a host of ideas necessary to imbed value focus in how initiatives are planned and executed

8.1 VALUE-DRIVEN WAY OF WORKING

This topic must be a key focus for your growth, whether you are a part of a PMO, playing a program/project coaching role, and/or working on programs/projects within the PMO. Value connects us. Having a clear link to the strategy behind our actions or our programs is key to our success both personally and professionally.

Program Management: Going Beyond Project Management to Enable Value-Driven Change, First Edition. Al Zeitoun.

Key Learnings

- Review the value-driven way of working with a failure in a pharmaceutical company program delivery – if you don't keep the value as a central component
- Exploring the changing way of working and its impact on program managers skills
- Using an example of the increasing trust in the PMO when it operates more strategically and at the enterprise level – directly tie this to operating more strategically

Reviewing the PMI's PMBOK Guide – Seventh Edition, we can see the importance of senior leadership setting the tone of focus on outcomes and benefits. In order to have a meaningful flow of performance information, that can be analyzed, it is critical to review the proper connections between strategy, portfolios, programs/projects, and ultimately operations. Continual feedback is essential to ensure that the PMO is driving the needed connection to the changing program ecosystem convictions.

- End-to-end view of the portfolio
- Cascade of benefits from portfolios into the programs in a continuous link
- Remember the difference between outcomes and outputs – need meaningful outcomes
- Continuous further adjustments in future programs
- Keeping an eye on the True North

8.1.1 Value-driven Way of Working – Pharma Case Example

This is a case previously mentioned, where one could miss the boat due to the high focus on activities and outputs and miss the view of true stakeholders' value and be distracted by the noise in the system.

- Strategic
  - Continual link across portfolio (multiple number of programs going on at the same time)
- Outcomes
  - Mindset shift and supporting processes (team could easily miss that so we need to keep connection to the changing needs of the stakeholders)
- Information
  - Integrated and cascading flow
- Value can take so many forms
- Products, services, and results
- Transformations
- Sustaining and innovating ways of working (endlessly as a continuous exercise)
- Enablers could be vast:
  - Charters (good design is value-centric)
  - Roadmaps (choosing the right horizons and what to accomplish in each of those as in previous lessons)
  - Transformative leadership and sponsoring (very important to review and understand the role of the sponsor, remind ourselves to bring back value in the driving seat of our focus)

8.1.2 Changing Skills

Combining the people and the technology sides, e.g. the recent shift in PMI's talent triangle with the focus on business acumen, power skills, and ways of working. It is critical to recognize that the skills for the future will continue to evolve fast. The more technology becomes a stronger enabler, the more the human skills have to adjust to match the jobs and work of the future.

Reminders of Skills

- Empathy
- Deep Listening – ways of listening
- Agile Working – enough degree of adaptability and resilience
- Stakeholders-centricity – entire ecosystem
- Digital fluency
- Brainstorming
- Design thinking – mindset shift to closeness to customers and key program stakeholders

8.1.3 PMO Operating Strategically

Controlling and policing have been changing and are not central to what governance is concerned with anymore. Future organizations see the PMO as a strategic arm. Executives expect the PMO to help them operate strategically, forecast obstacles, and focus on efficiencies that support growth and achievement of business outcomes.

- PMO as a Strategic Unit:
  - Leadership in governing with excellence (does it enhance the program teams' ability to produce results)
  - Integrating across internal silos and external partners (PMOs should break these down using trust as the future currency for transformation)
  - Should the PMO consider scaling using a PMO-as-a Service model to allow wider and more efficient access to its services?
- Role of Trust:
  - Support for executives' top priorities (making it clear you are helping their strategic agenda advancement)
  - Driving strategic priorities (right programs and outcomes toward the value needed)

The So What:

- PMOs will shift
- PMOs are strategic entrusted units
- They drive focus toward value
- They ensure we are investing in the skills that matter to the future of work
- Highest focus is on making sure the world around us gets the right strategic benefits aspired

Review Questions

Parentheses ( ) are used for Multiple Choice, when one answer is correct. Brackets [ ] are used for Multiple Answer, when many answers are correct.

_____ is a critical shift in how future organizations' way of working matures in the next decade.

( ) Focus on operations
( ) Outcomes focus
( ) Deliverables integration
( ) Maturing processes

Which of the following are examples of the skills in focus as program managers adapt to the changed WoW?

Choose all that apply

[ ] Directing
[ ] Digital fluency
[ ] Performance measurement
[ ] Deep listening

Which of the following is an indication of the maturing strategic role of the PMO?

( ) Leadership in governing with excellence
( ) Conducting more training sessions
( ) Happy team members
( ) Being the first to jump into the technical program aspects

8.2 THE VALUE MINDSET

It is easy to talk about the importance of a value mindset, yet is harder to truly shift. This shift needs to be in multidimensions that cover the culture, processes, ways of working, behaviors, and ultimately execution approaches and results. This need to be thought of and managed as a critical change management exercise. Program managers need to learn and practice ways for creating that shift in order to get ready to achieve programs' value more consistently.

Key Learnings

A few key reminders and some new things to focus on to shift the attention toward value delivery:

- The worst scenario – focusing on busy program work (what are the consequences if I am just busy and heads down in my programs?)
- Exploring cultural attributes to support building a value mindset (Peter Drucker's culture east strategy for breakfast and how to get the right support)
- Linking to excellence and exploring the EFQM Model linkages to value mindset (some of the attributes that confirm the standards for us to benefit from)
- Value mindset allows the PMO to become the strategic arm of the business (what does the future look like when we implement those shifts?)

8.2.1 Busy Program Work

One of the likely big pain points/questions, affecting focus on value is: Are we spending our time in the right areas? Noise, internal politics, and distraction all around us, so we need to be present and conscious of these challenges.

- What are key lessons if extreme activity focus still dominates way of working in organizations today? (The daily need to review and reflect on this).
- What will happen if program team is super busy delivering and customer is unhappy? Customer Experience (CX) is vital and the only way to excel in our operations' focus.
- How do programs measure their performance with a proper view of value? Consciousness and awareness coupled with emotional intelligence and being present of our actions' impact.

8.2.2 The Decision to Building a Value Mindset

As indicated in the PMI's PMBOK Guide – Seventh Edition, building a value mindset, is like being at a crossroads:

- PMI defines value as the ability to continually evaluate project/program alignment to business objectives and the intended benefits/value (theoretically easy yet practically, life happens so we need to cascade well across teams).
- This decision is a mindset culture shift, yet seems natural to the sustainable existence of tomorrow's organizations (need to ensure the entire orchestra (organization) is tied to this changing focus).

8.2.3 EFQM Model Linkages to Value Mindset

Reminders as reflected in Figure 8.1:

- At the end of the day, purpose brings us back to value
- Agility helps our ability to adjust and step away from pure activity focus
- Investing in the creativity needed for success
- Diversity creates such a key dialogue needed for value
- Include even the ones who you are least inclined to include
- Innovation is a muscle that also support sustainability agendas and missions like the UNs
- Ensuring that the PMO is embedding the advocated skills critical for success
- Stretch in offering of skills for wider transformation results, e.g. PMaaS (Program Management-as-a-Service)

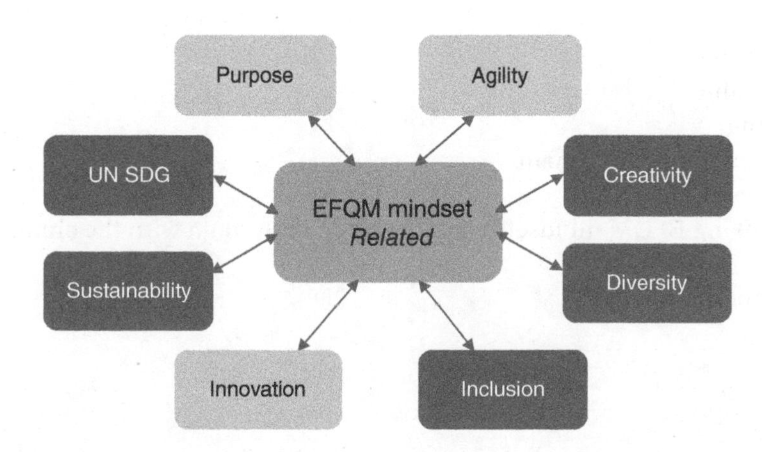

FIGURE 8.1 EFQM and Mindset
Source: Adaption from the European Foundation for Quality Management (EFQM)

8.2.4 Value Mindset and the Strategic PMO

PMOs in the future are shifting to true ***strategic offices***. These entities are no longer taking a back seat or are expected to be just supporting or reporting units. There is a heightened demand for PMOs to drive impact in the delivery of strategy. For this impact to happen, the PMO should question every program, project, and action, and whether they map to value. The idea is to guide the organization and its stakeholders to see work with a different set of lenses:

- PMOs of command and control will disappear (history)
- Where should the PMO sit to drive value and what are the PMO formats of the future? (This is the key shift – asking the question is there a difference when the PMO is not there tomorrow?)
- Changing Metrics focus (inputs to outputs) and the energy around Objectives and Key Results (OKRs) and their use in driving strategic focus (designing and investing in those and having a team that is central to driving that)
- The future of PMOs is strategic agility (This is where the future PMOs will focus and create shifts in strategy interactions very differently from how it is done today)

The So What:

- Remember the practical learning point of where you spend your time
- Program teams make a decision to make value front and center
- Every dialogue, every planning exercise shift to this type of discussion
- Requires a different type of a dialogue
- None is able to do that shift better than the PMO
- The fork in the road is a conscious decision of altering focus

Review Questions

Parentheses ( ) are used for Multiple Choice, when one answer is correct. Brackets [ ] are used for Multiple Answer, when many answers are correct.

_____ is a critical consequence of a program culture that prioritizes extreme activity and keeping teams busy.

( ) Enhanced collaboration
( ) Ineffective view of value
( ) Openness to thinking
( ) Strategic view of program management

Which of the following EFQM mindset-related points closely align with the changing focus on value in future programs?

Choose all that apply

[ ] Sustainability
[ ] Purpose
[ ] Trust
[ ] Innovation

Which of the following is an example of a metric that reflects clear focus on benefits?

( ) Staying within 10% of the budget thresholds
( ) Increasing the program team's creative ideas inclusion rate by 20%
( ) Completing all the scope deliverables on time
( ) Being the best in class in responsible AI

8.3 BENEFITS MANAGEMENT MATTERS

It has become clear that one can't talk program management without discussing and confirming the approach for achieving program benefits. It is fundamental to build benefits into the program's frameworks, practices, collaboration across program teams, and across stakeholders. The PMO has the ownership to ensure benefits are central to what we do and how we achieve that.

Key Learnings

- Is the program manager suited to drive benefits management? (Our ability to be as candid as we can and what the PM needs to change)
- Defining the critical importance of benefits management to transforming the business (why is it so critical and ensuring the understanding across stakeholders)
- Establishing clear understanding of benefits management in the consistent actions of the PMO (including all the parties who are interacting with the PMO and programs on an ongoing basis)

8.3.1 PMO and Driving Benefits Management

Figure 8.2 highlights five critical ingredients for achieving a focused drive toward benefits management.

Reminders:

- PMO Sets the stage for program managers to drive benefits
- ***Business acumen*** is one of the three sides of PMI's talent triangle – asking the questions and connecting to the business leaders
- Time and again, we talk about the importance of looking ***end-to-end*** as in Program Life cycle Management (PLM)

FIGURE 8.2 PMO and Benefits

- Always ***rethink*** what success looks like
- Also making sure the culture of ***ownership*** is supported and exhibited by the action and behaviors of the program managers
- Maintaining the line of sight with the executives via the continuous interactions and having a ***continual sensing*** of what is important to them, including talking with their executive assistants

8.3.2 Transforming the Business

The three building blocks for transforming the business remain process, infrastructure and technology, and people:

- Benefits strategic dialogues at center of transformation
- Reversing trend of transformation failures with value-focus (learn from just the pure focus on speed versus what the full value is)
- Refreshed stakeholders engaging (need for energizing around the change)
- Adjusting run-change mix (the future in the project economy is about adjusting the mix and a much more shift to programs instead of the classic Chief Operating Officer, COO, and more into a Chief Program Officer)

8.3.3 The PMO and Consistent Benefits Management

- As shown in Figure 8.3, the benefits understanding starts with the PMO being clear on communicating its purpose
- Classic PMO leadership is shifting away from being control-driven to more servant-leadership like
- Customers' needs remain at the center
- Daily interaction around the business needs
- The ways executives delegate and operate have to shift
- More and more the data and insights that can turn us back to benefits are critical for our success

The So What:

- We have got to make sure that the PMs and PMOs are invested in the shifts toward benefits matter
- This is shown in the types of programs we get involved in
- The transformations of the future are value-centric
- PMs are realizing the shift that is taking place toward more of the change the business initiatives becoming almost 80% of what we do
- Making sure the focus on benefits management is highlighted in the focus and support of the executives and the way by which they lead

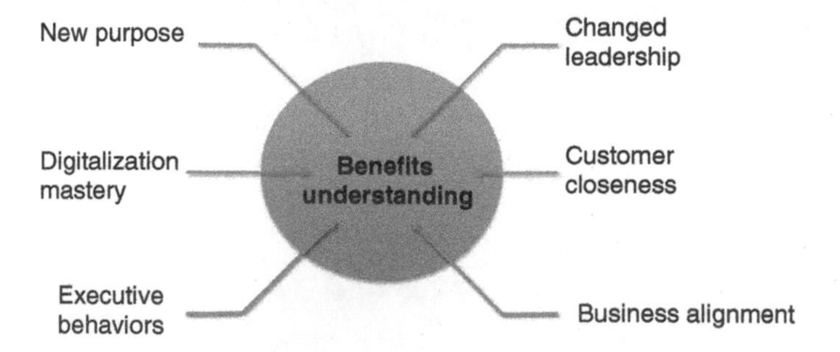

FIGURE 8.3 Benefits Understanding

Review Questions

Parentheses ( ) are used for Multiple Choice, when one answer is correct. Brackets [ ] are used for Multiple Answer, when many answers are correct.

_____ is an example of what qualifies a program manager most to drive a benefits management focus.

( ) Anchor and maintain one view of success
( ) Exhibiting a strong business acumen
( ) Deep IOT expertise
( ) Allowing program teams full creativity

Which of the following shows the critical importance of benefits management to transforming the business?

Choose all that apply

[ ] Adjusting the business portfolio's run-change mix
[ ] Focusing transformation missions on applying more digital solutions
[ ] Agility in strategic dialogues
[ ] Minimizing stakeholders' interferences with programs

Which of the following is an indicator of the PMO's ability to understand the critical importance of benefits?

( ) Maintain a close focus on budget
( ) Ensuring distance from customers for objectivity
( ) Standardizing the use of Excel in program templates
( ) Using coaching to alter executives' behaviors

8.4 INITIATIVES SUCCESS

Success of programs hinges on our ability to think for a change. We usually don't question things and think the executives know exactly what messages are we cascading. We should know better that with the changes into agility and how we work today, that intentional communication of purpose and approach is critical. We have got to make sure we have the ability and mechanics to think about what success truly looks like. Refreshing one's attention as a program manager for what is truly an affective plan is also of utmost importance.

Key Learnings

- Understanding the importance of being value-focused in program solutions
- The BAH secret sauce of empowering staff to turn focus on program's value and highlighting the people element of programs success (across clients and governments around the world)
- Identifying the role of establishing continual program connection between the head and the heart to increase initiatives success (the head, heart, and hand tool we discussed before to help us accomplish this)

8.4.1 Value-Focus in Program Solutions

When we view the life cycle of a portfolio of initiatives, we see that its components are through parts of that life cycle achieving their own outcomes while potentially contributing to the joint overall integrated outcomes of the portfolio. For this to happen well and achieve the most value:

- The value needs to be determined at the portfolio level and cascades into the programs and projects
- Cadence and dependencies have to be discussed
- Iterative rethinking of success (the flexibility in moving programs and creating overlaps)
- Focus on impact (how do we ensure all the components collectively create the value)
- Stakeholders' engagement (using all the tools of analysis and engaging we studied before)
- Product mindset (bringing this to the mix with product teams and product manager capabilities into the role of program managers to bring us to value and rethink the mix)

8.4.2 The People Element in Program Value

- People are behind the scene in every aspect of this success
- Context clarity matters (user cases, storytelling, business purpose)
- Leaders excelling in defining purpose (nothing is more important than the leaders standing in front of their troops and identifying expected value)
- Inclusive leadership approach (having the right experts, segment leaders, and other key players who know what is needed for whatever business model: B-2-B, B-2-C, or even B-2-B-2-C)
- Although a technology organization, it is a people business

8.4.3 The Head and the Heart

- Figure 8.4 shows the sequencing and the continual elaboration of this important connection. Articulate with the head, engineering mindset, technology, as a nice thread
- The heart is critical for energizing action (nothing is as successful as when the hearts are ignited)
- Continual process between head and heart (articulate something different to drive the hearts when needed)
- Maturing business and achieving successes are about the consistency in practices of excellent behaviors

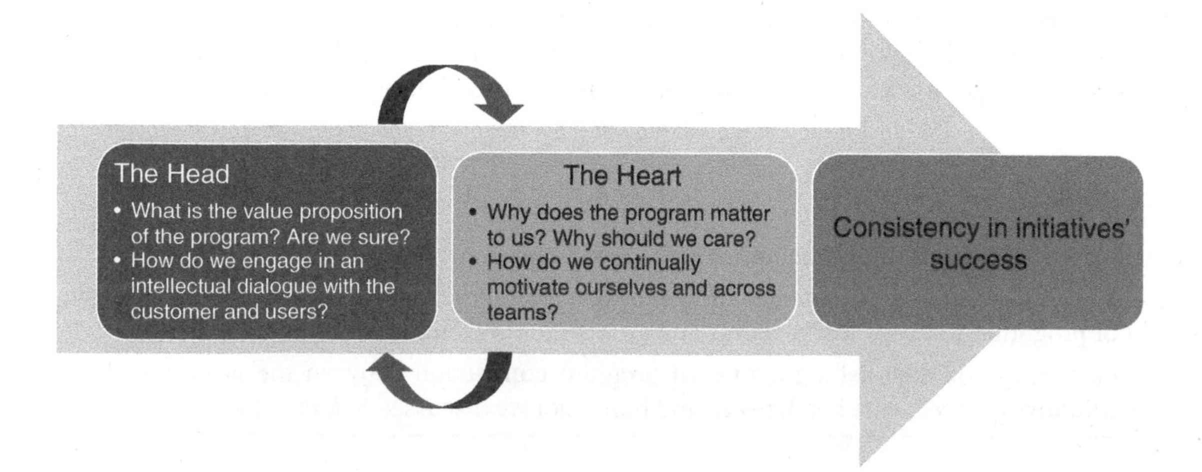

FIGURE 8.4 Continual Head–Heart Connection

The So What:

- Challenge success
- Make sure the leaders know where they have the biggest impact
- Ensuring including diverse views
- Look holistically across the program
- Interactions across programs
- Catch early on what success alterations might be needed
- Let's not hesitate to bring the tough questions, not leaving any elephants in the room
- The PMO's ability to drive consistent dialogue creates a strong head–heart alignment

Review Questions

Parentheses ( ) are used for Multiple Choice, when one answer is correct. Brackets [ ] are used for Multiple Answer, when many answers are correct.

______ is not an example of a proper focus on value in a given program's solution delivery.

( ) Developing a product mindset.
( ) Focus on inputs metrics.
( ) Iterative thinking of program success.
( ) Portfolio breakdown into components focused on impact.

Which of the following are good examples of incorporating people in the delivery of programs' value? Choose all that apply

[ ] Confirming clarity of a program's purpose first
[ ] Jump into getting work done fast
[ ] Involving a small circle in decision-making
[ ] Taking time to define definition of done

Which of the following shows that the program team's hearts are engaged in driving toward success?

( ) A nicely written program charter
( ) Minimal team's interactions
( ) Energy and motivation for program's work
( ) The program manager's detailed program plan

8.5 IMBEDDING THE VALUE FOCUS

PMOs need to make it a priority to drive the creation of a mindset, tilting toward value, and what that shift requires. Answering the question: What are the additional tools/enablers to ensure the planning, execution, and the changing of the attention to value all comes together? Anytime we talk about a change in behavior, it is hard. This PMO effort is about high change management focus. Changing how the program teams operate and where do they spend their time are a sign of succeeding in that mindset shift.

Key Learnings

- The Booz Allen Hamilton example of consistently driving the instilling of PMO value across clients' organizations (making sure as consultants, they transition value to the customer)
- Positioning value focus in the daily drive of the program manager (behaviors have to change and thus consistent repetition is needed)
- The Program Team Charter example as a clear way of capturing value focus (importance of clear roles and responsibilities and that the team can tailor their stories and work toward value)
- Tailoring the value-focus stories to inspire successful programs outcomes (storytelling is very important both in the communication and in the entire program approach)

8.5.1 Driving PMO Value

- The PMOs' future value hinges on their ability to integrate and unify (align across lines of business, consistent ways of going to market, working with clients need that secret sauce of delegating to protect your brand in the consistency of operating and focus)
- If program teams come across as different organizations, the business brand is at risk
- Seamless organizational flow across market segments and clients creates a connected success story and builds learning muscles (more joint success is what we are after more than the individual successes)
- The leadership of the future requires a renewed commitment: Our vertical numbers matter, yet ultimate success is in the horizontal spaces within and across ecosystem partners (humbleness in the learning and focus on catching the gaps in the white spaces and linking with an ecosystem mindset including our partners and how much are they part of our understanding of value)

8.5.2 Daily Drive of PM

Consider the following daily drive creation questions:

- Are we focused on the rights things? (To create the right routines with a portfolio mindset)
- Are we reviewing Enterprise Risks? (How can I expand that to the enterprise level)
- Are critical dialogues taking place? (The idea that the PMO is the go-to place to drive the right dialogues)
- Is transparency dominating our team's communications? (The volume, the nature, and the intentionality of communications)

8.5.3 Team Charter and Value Focus

One could also revisit the use of the program team's charter from an angle of focus on value:

- The purpose contract that sets the stage for how we work as a team (anchoring the commitment, an evident way of committing the team to value)
- The clear roles and the Why Statements of team members (e.g. 16 personalities tool to develop a joint view about the team)
- The rules of engagement and ways of getting back to value (if not, then why and why not)

8.5.4 Tailoring Stories and Outcomes

The PMO could articulate and tailor the change journey stories in a way that highlights Outcomes' Excellence:

- A program's strategic score (e.g. objectives and key results [OKRs] to map upward to an overarching strategic score)
- Hearts' engaging information radiators (making sure tools are connected to driving the behaviors we want to have)
- Value performance charts (Burnup, Burndown, Combined) – classic ways to better tell the story about performance
- Using a business canvas (having a visual way to the story as a meaningful way of focusing on outcomes)

The So What:

- The routines and consistency that the PMO creates
- The ability to go back to tools, such as the Team Charter to anchor commitment to value
- To deliver excellence, bring things back to the connection to the team and ecosystem partners
- A multitude of things that come together as an organization to plan, execute, and pull the right stakeholders together, with the proper degree of fluidity
- Creating the hybrid future

Review Questions

Parentheses ( ) are used for Multiple Choice, when one answer is correct. Brackets [ ] are used for Multiple Answer, when many answers are correct.

_____ is a critical ingredient that allows business segments and their clients to focus on achieving value

( ) Designing performance measures within business buckets
( ) Sharing success stories across verticals and partners
( ) Focus on relentless control
( ) Allowing each vertical to do what most fits each of them

What are the top benefits of having a team charter in today's VUCA environment? Choose all that apply

[ ] Minimizes change
[ ] Captures the rules of engagement
[ ] Ensures clear scope
[ ] Becomes a value contract

Why is it critical that the program manager tailors the value-focus stories to the specific nature of the team?

( ) She/he enjoys telling stories
( ) Different stages of maturity and emotional connection
( ) Keeping an eye on governing is a must
( ) Earned value management applies well

8.6 THE HYBRID WAY OF WORKING

It is important to recognize and understand the movements to a hybrid way of working and what does that look like. PMOs can take the lead on driving examples for how to co-create the hybrid program approach. Program managers should also learn how to insert the right degree of fitting agile practices. Working with the PMO, program teams explore examples and strategies for methodical shift to a hybrid organization. With more experimenting, we develop an appreciation for applying hybrid practices in industries, where traditionally this could not be thought of, like in construction.

Key Learnings

- Showcasing the Dubai government transformation
  - a leading example of hybrid working (the nuclear program, Expo 2020, and other events that were disrupted by world events)
- Resilience capabilities (to address such mega programs)
  - a few pandemic examples
- How collaboration dominates the way of working?
  - programs examples
- Understanding organizational viruses that impede agile working

8.6.1 Why the Shift?

- Adapting and setting the stage for how we work differently with technology and other major way of working shifts
- Experienced that in pandemics and changing world dynamics
- What opportunities this creates that we have to keep central to the future way of working as program managers
- There is going to be some shifts that we want to embed in how we continue to work in the future

8.6.2 Hybrid Working Transformation Example – Dubai[1]

1. Six Strategic Growth Pillars for Dubai Emirate (Emirate Charter put together to look at creating a focus for the organization, this case being an entire Emirate)
2. Ensure transparency and sense of responsibility in decision-making (how this dominates the future of the Emirate)
3. New transformations to continually overcome emerging challenges (proper collaboration across the leaders and using data and analytics to ensure the work will be properly accomplished)
4. Better Future for Next Generations
5. World's Most Livable City (Is this possibly the North Star and how this gets an Emirate like Dubai to learn from the past and tackle consistently the demands of future programs)

TIP

Having a transformative and empowering vision, coupled with clear roles and responsibilities could make the shift to a hybrid way of working most effective.

[1] Adopted from the Dubai Charter – January 4, 2020

8.6.3 Resilience Capabilities

Pandemic Example - as highlighted in Figure 8.5

1. Design and engineering (simulation, speed of decisions) – focus on matters most, in this case human life
2. Products (new lines utilizing existing capabilities) – almost overnight tilts from a current DNA to serve unique needed products and services
3. Ways of working (rethought collaboration and silos breakdown) – like in the cases captured in the PM Next Gen book about Brazilian organizations that managed overnight to turn the focus onto products that makes sense
4. Innovation centers (team of teams) – thinking programs and project as labs to seamlessly integrate and experiment immediately and get feedback from users and customers and not just rely on government and policies but also create the right delegation and autonomy muscles

Collaborative Way of Working

- This is a must
- Concept boards (e.g. Miro)
- Social check-ins (use the collaboration vehicles to connect us across the Globe and across complexities, very valuable into the future)
- Rotating leadership (great way of building the activation of motivation of the team and future generations)
- Service focus (seeing the linkages to the team with relentless questioning)
- Continual purpose questioning

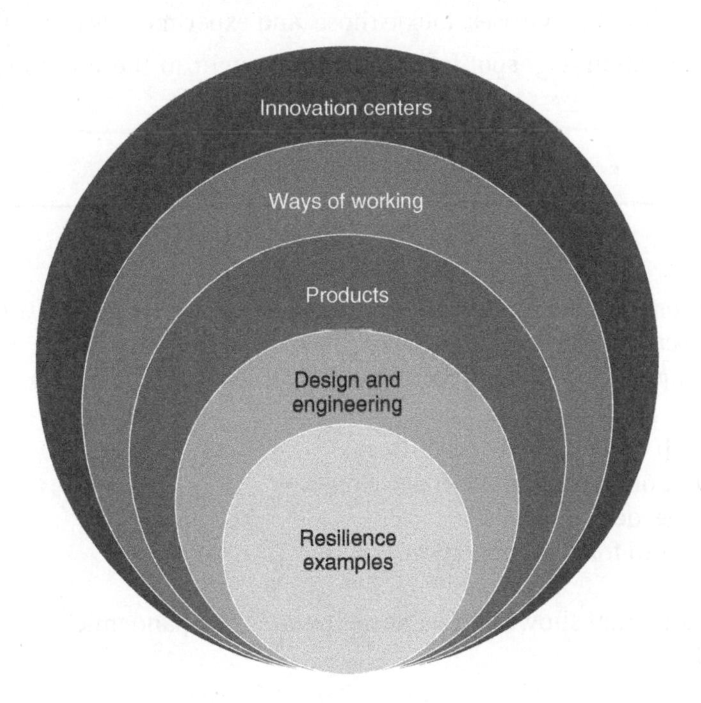

FIGURE 8.5 Pandemic Example

8.6.4 Agility Limiting Organizational Viruses

Having discussed organizational viruses earlier, we know that they could be limiting the program teams and their leaders' abilities to drive transformative change. These viruses should gradually be tackled and the ways of working get adjusted in order to sustain growth and progress:

- There continues to be a ton of organizational viruses, e.g. senseless processes, the silos, the components of the portfolio will benefit when resistance melts
- Cultural resistance
  - Inability to implement hybrid strategies
- Prevailing mindsets
  - Unwillingness to experiment
- Imaginary need for control (leave this up to the team with some minimal degree of control)
  - Destroying autonomy needed for hybrid testing
- Slow and weak decision-making muscles (a classic organizational virus that should be broken down with data and encouraging proper behaviors)
  - Creating bottlenecks to transformation

The So What:

- Setting the stage for the week with the hybrid way of working
- Organizations that use their purpose well can figure out the right degree of alignment, delegation, and autonomy
- Having clarity of organizational viruses, tackle those, and experiment is critical
- The idea of cross-pollination, e.g. spotify, need to be brought to the mix to ensure the program team is effective

Review Questions

Parentheses ( ) are used for Multiple Choice, when one answer is correct. Brackets [ ] are used for Multiple Answer, when many answers are correct.

Which of the following reflects Dubai Charter's commitment to Hybrid work?

( ) Leave growth agenda up to individual committees
( ) Transparency and sense of responsibility in decisions
( ) Transform only when needed
( ) Stay with what has worked for this generation

What are good examples that showcase resilience in times of a pandemic?
Choose all that apply.

[ ] Ability to reprioritize the mission and create products' adjacencies
[ ] Decide on ways of working that adapt to global conditions
[ ] Become popular with government agencies
[ ] Rethink engineering models between the physical and virtual worlds

What organizational viruses could limit the implementation of a hybrid way of working?

( ) Experimentation mindset
( ) Dynamic culture
( ) Comfort with autonomy
( ) Was not invented here

8.7 COCREATING THE PROGRAM APPROACH

One of the differentiating capabilities of the future is cocreating. It is one of those real connectors to customers, users, and across teams. The future depends on cocreating to create buy-in. This is how you could demonstrate care about the customers and the folks on your team. It is very important to look at a set of data around agile and how those practices could make the program teams more effective.

Key Learnings

- The right program approaches
  - Rethinking what that looks like
- Program approach design flexibility and correlated program success
- Exploring State of Agile Report Data:
  - Programs success shifted
  - Cocreation in future company cultures
- Consistency in experimenting with agile practices and process (to have the right agile practices brought into the mix)

8.7.1 The Right Program Approach

- Not one size fits all (remember the importance of that in designing your program roadmap as we discussed before)
- Not about skipping proper planning (hybrid does not mean chaos, it is about finding the right balanced decision-making)
- Design ingredients:
  - Collaborative (right inputs and people)
  - Principles based (we have seen the shift in many of the program/project standards)
  - Human-centered (is the focus on the people and stakeholders)
  - Digitally enabled (I should not miss out on the value of making the most use of digital in connecting the virtual and wider ecosystem)

8.7.2 Program Approach Design Flexibility

Flexibility and the UAE Program Success

- In the energy program of the country
- Rethinking contribution of each organizational team (starting from a blank sheet of paper)

- Metrics that drive horizontal behavior (rethinking in order to excel in the horizontal integration)
- Highlighting both successes and failures' learnings (how we meet, focus, and conduct reviews)
- Providing leadership opportunities for local workforce (finding opportunities to grow the local leadership and the Next Gen team members)
- Use any selected approach as a guide only (use flexibility to drive the impact of the approach you choose)

8.7.3 State of Agile Report Data[2]

Agile Adoption: The data were consistent around value

- ***Success = Value Delivered***
- Enhanced changing priorities management (getting to a place where prioritization is seamless)
- Transformation for customer (we do this in order for customers to have a better-connected experience)
- New ways of thinking (the movement toward agility and finding the right degree of a hybrid approach to empower the program teams)

8.7.4 Consistency of Experimenting

- Success = Consistency of valuable patterns
- The voice of the business is reflected in cocreation (am I listening to the business and the interactions with the customers?)
- PMOs create experimentation culture (allowing consistency to prevail)
  - supported by new and adaptive leadership models
  - encourages and coaches' collaborative behaviors
  - learning stretch opportunities (learning enables us to scale and see things we have not envisioned before)

TIP

Having cocreation in focus, helps the program manager connect well, and enables the PMO to develop effective adaptive cultures and leadership models.

The So What:

- Cocreation enables success
- The amount of experimentation creates the trysting partner role for the PMO
- As we look at success stories across organizations, we extract the learning that expedites the movement toward the hybrid way of working
- Enjoying the successful outcomes of that shift

[2]Adopted from the State of Agile Report 2022 – https://www.scruminc.com/2022-state-agile-report-takeaways/

Review Questions

Parentheses ( ) are used for Multiple Choice, when one answer is correct. Brackets [ ] are used for Multiple Answer, when many answers are correct.

Which of the following is a healthy ingredient in the design of a fitting dynamic program approach?

( ) Rolling out one size for the organization
( ) Creating a comprehensive list of templates
( ) Full automation is a must
( ) Principles based

Which of the following elements are examples of how success in the future is linked to hybrid ways of working?

Choose all that apply

[ ] Success = timely scope delivery
[ ] Effectiveness in managing changing priorities
[ ] Control of customer requirements
[ ] New ways of thinking

Why is the future PMO critical in creating a culture of experimentation in program work?

( ) It rewards central leadership
( ) It drives all key decisions
( ) It coaches collaborative behaviors
( ) It ensures once a program approach is created, it sticks

8.8 VALUE OF FLEXIBLE DELIVERY

The shift to hybrid work means value needs to be demonstrated. This is a selling aspect of the PMO strategic role. The key takeaway in achieving flexible delivery, will be outcomes integration success. PMOs help us look at patterns of successful transformations to repeat those successes across other major change programs.

Key Learnings

- The worst scenario – not being flexible (that is a no go)
- Linkages across program metrics to highlight the importance of flexible delivery (focus on flexibility in delivery)
- Connection between proper value focus and the 30% of transformation programs that succeed
- Examples to understand different delivery methods (or at least refresh that understanding as we had covered across prior lessons on this)

8.8.1 Worst Scenario

Figure 8.6 reminds us of the unique nature of constraints in programs and what is takes to achieve program value. There is a need for the PMO to choose the right handing mechanisms for uncertainty and complexity, while the PMO charts the change course ahead:

- Missing flexibility = win the battle and lose the war (would be a nightmare)
- Change makers of the future understand this triangle (change scientists)

- Keeping our eyes on program value ensures we win the war
- Unhappy customers (tilting enough toward their needs as mentioned in many industries example before)
- Inability to see opportunities side of risks

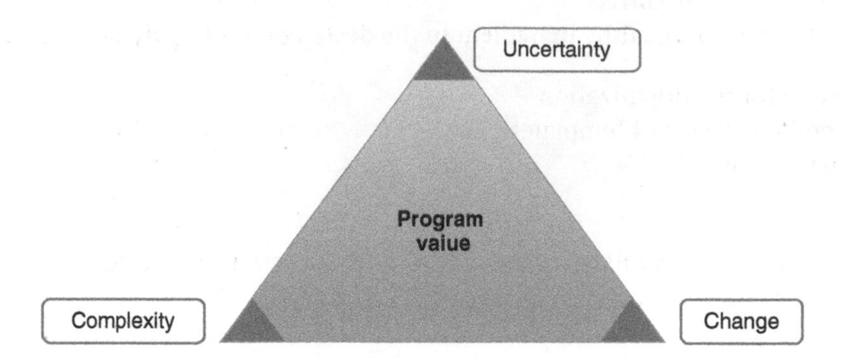

FIGURE 8.6 Balancing Program Constrains to Achieve Value

8.8.2 Linkages Across Program Metrics

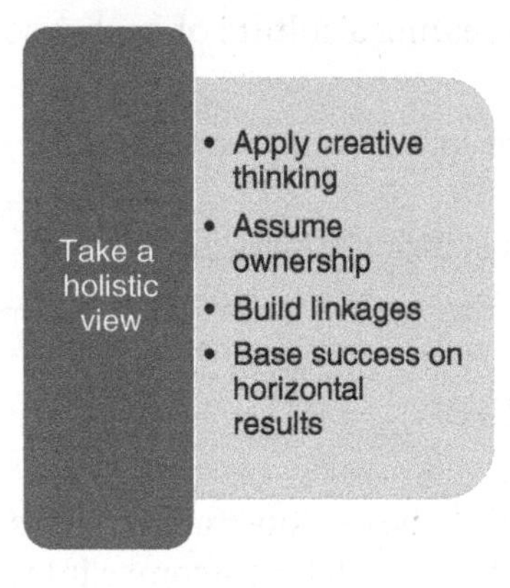

- Visualize across verticals
- Apply consistent reviews
- Unlock potential with stretch goals
- Decentralize decision-making
- Organize around value
  - *Metrics*: tell me how am I measured, I will tell you how I will perform
  - Worst thing would be shooting ourselves in the foot by picking contradicting metrics
  - Using mind maps to show connection between metrics
  - Need continual dialogue about the metrics to ensure decisions are not hindered

8.8.3 Connection to the Successful 30% Transformation Initiatives

- Figure 8.7 summarizes a number of critical ingredients for transformation success: A small percentage of transformation programs succeed
- Partnerships mindset helps the movement across silos inside and outside the organization
- Removing obstacles shows the right team responsiveness

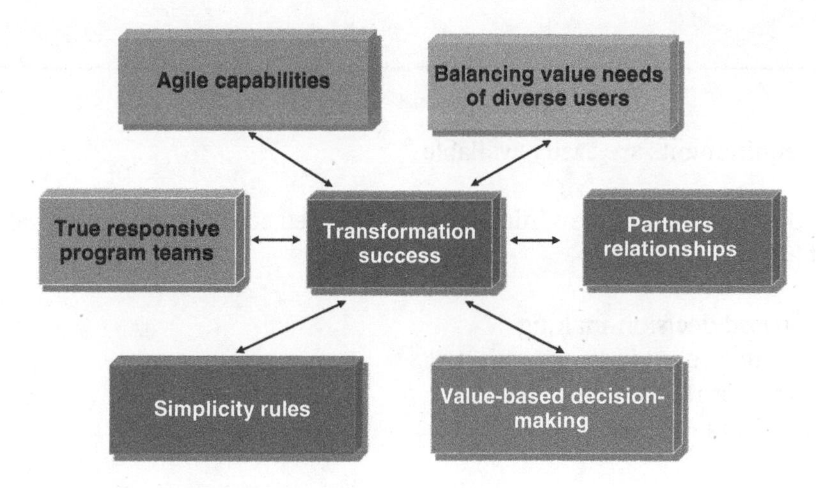

FIGURE 8.7 Transformation Success Building Blocks

8.8.4 Different Delivery Methods

- Every organization has to find its right fitting approach (tailoring)
- Add more details as uncertainty decreases (what is the right degree?)
- Critical ways of designing:
  - A pharma organization that rewards the hybrid behaviors (one would see immediate positive impact)
  - A global organization's portfolio that struggles to change fast enough (center of excellence can support this global behavioral shift)
- Build on organizational agility principles (matters most)
- Customize for best-fitting program delivery method that matches the state of organizational maturity (right degree that matches the organizational maturity by engaging the leaders and other stakeholders in the dialogues)

The So What:

- Most important to remember is the horizontal integration, e.g. across metrics
- Rethink our metrics
- It is about fit, readiness, and making sure we are not limited by a rigid view of success
- Ready to operate at the right degree of flexibility needed for future programs

Review Questions

Parentheses ( ) are used for Multiple Choice, when one answer is correct. Brackets [ ] are used for Multiple Answer, when many answers are correct.

When the worst scenario exists and program teams are inflexible, the following could happen:

( ) Program managers become more valuable
( ) In most cases, program teams miss responding with speed to changing clients' true needs

(continued)

(continued)

() We lose battles
() Comprehensive requirements are easily available

How does the 30% of transformation initiatives that succeed confirm the proper value focus? Choose all that apply.

[ ] They apply value-based decision-making
[ ] Base capabilities on one select industry standard
[ ] Use simplicity to remain responsive to users
[ ] Ask executives to remain uninvolved

Which of the following gives assurance that the delivery method will adapt to a changing client environment?

() Addressing the needs of executive sponsor
() Increase documentation
() Rolling out changes in line with increasing maturity
() Staying connected on social media

8.9 PROGRAM LIFE CYCLE CHOICES

Many program managers think they don't have a choice across the program lifecycle. Part of that assumption is that the program choices are just cascaded to the team from the executives or the customers. Program managers need to reconsider the roles and importance of autonomy and the tailoring for fit.

Key Learnings

- Autonomy and the tailoring of a program life cycle choice
- PMO leading the movement to insert the right degree of agile (think of this as a transformation exercise)
- Scaling and the methodical shift to a hybrid organization (going back to the Bosch story)
- Applications of hybrid work across industries (even in the construction industry, contracting, reality it does work there too)

8.9.1 Autonomy and Tailoring

Autonomy Is Not the Only Answer, As We Need to Consider As We Tailor Our Choice

- Having a clear vision and North Star (if missing, we struggle)
- Build the program life cycle around value delivery (North Star provides the linkages to value)
- Think about the program's complexity (this determines how far we can go)
- Consider the degree of technology disruption in the mix (market, customers, and technology)
- Test the risk appetitive of the program leadership (this goes a long way in deciding what exactly I can do especially in regards to scaling)
- Find the most effective solution for global inclusivity

8.9.2 PMO and Leading the Agile Practice

Key point on the PMO's agenda

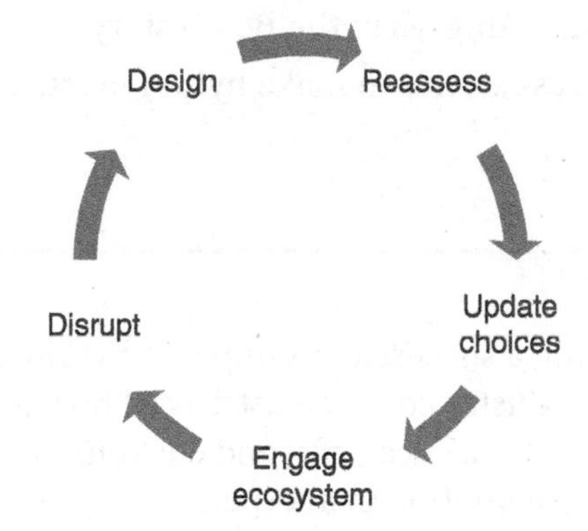

- PMOs are focused on strategic agility (flexibility by which I can adjust my strategy – ease and speed)
- The role of the PMO as the change engine (adjusting fast)
- Continuous improvement stance (how the PMO creates the adaptability in the way of working especially in the case of strategy)

8.9.3 Shifting to a Hybrid Organization

- Business acumen (such an important component of the talent triangle)
- Accountability driven by the Bosch board (white board, rolling up the sleeves, acting as product owners)
- Product ownership and its cascade across lines of the business
- Strong sponsorship (commitment)
- Consistent engagement infrastructure (consistent engaging)
- Always disrupt mindset (letting go of some of the control to get to a hybrid organization)

8.9.4 Hybrid Across Industries

- Figure 8.8 demonstrated a few key points about the hybrid way of working: The hybrid way of working works across the board
- Even how we offer educational programs
- Committing constriction contracts to a price that is fixed and then leaving a good portion that has flexibility requiring a high degree of trust by the customer
- Got to have hybrid at all levels, including governments and policymaking bodies

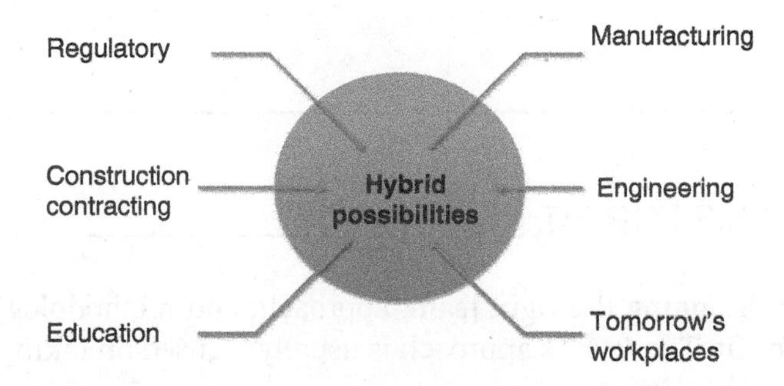

FIGURE 8.8 Hybrid Across Industries

The So What:

- The PMO owns the agile practices
- Got to be relentless about experimenting, as in the Bosch story
- Create enough example and success stories to make hybrid practices contagious enough across all areas that the PMO touches

TIP

Having a successful example of hybrid in one industry could be used as a "how to guide" for a successful and customized implementation in another.

Review Questions

Parentheses ( ) are used for Multiple Choice, when one answer is correct. Brackets [ ] are used for Multiple Answer, when many answers are correct.

_____ is an example of a factor to consider in tailoring the most fitting program life cycle choice.

( ) Number of gates required
( ) Programs complexity
( ) Most challenging stakeholder
( ) Emotional intelligence

Which of the following prepares the PMO to lead the agile practices movement?
Choose all that apply

[ ] Centralized governance
[ ] Operating as the organizational change engine
[ ] Staying away from politics
[ ] Continuous improvement way of working

Which of the following shows an industry where hybrid approaches could be implemented?

( ) Education
( ) Construction contracting
( ) Manufacturing
( ) All of the above

8.10 ORGANIZING TEAMS FOR FIT

Fit is a fundamental word in designing the right team, approach, and methodology. The PMO should lead the orchestration of this design for fit. The PMO's approach is usually focused on taking baby steps for changing the organization to become more value-driven.

Key Learnings

- Maturing of matrixed way of executing programs (the need to comprehend how this works)
  - A PMI focus area
- The learning organization and exploring the right hybrid degree fit (authorize the team differently)
- Balancing the role of program leaders
  - Sensing the readiness for hybrid program work
- Creating the strategic mix of Waterfall and Agile (what is the right split?)
  - Is it a 50/50 split, or is it going to be a more fitting split? (Looking at complexity, readiness, and leadership)

8.10.1 The Matrixed Way

The PMO and the program manager own this horizontal way of working.

When the program manager operates as conductor, the focus is on organizing horizontal harmony.

PMI has been driving the importance of matrix way of executing projects and programs:

- Enablers for a strong matrix organization
- Closing the gap between strategy development and implementation (closing that gap hinges on the ability of breaking down the organizational silos)

8.10.2 The Learning Organization and the Right Fit

- Different muscles and capabilities
- Leaders have to continually practice new thinking models (e.g. the UAE example) and
- UAE Program example of cross sharing across three key PMOs (corporate, construction, operational readiness having to align)
- Program teams thrive when organization supports their learning:
  - Fear disappears and team focuses on exploring right fitting hybrid model fast
  - Authorized team members are more creative and create scale opportunities for cross pollination of ideas (cross pollination of ideas across PMOs and across the organization)

8.10.3 Balancing Role of Program Leaders

- Leaders show the way: crystal clear stories, connect the hearts, and entrust teams (matters the most for success and for team's dynamics)
- ***What does done look like***[3]: 5C's contribute to creating program teams that fit the situational complexity

[3]Adopted from Prowess Project – What does done look like? – https://prowessproject.com/what-does-done-look-like/

- Color (backdrop)
- Context (understanding our angle)
- Connective tissue (build on the matrixed way of working)
- Cost
- Consequence (helps with situational understanding well)

8.10.4 Creating the Right Strategic Mix

- Rethink where value fits best
- Ensure prioritization success potential (seamless across the business)
- Cancel unnecessary gates (if it does not bring value)
- Create spaces for massive collaboration
- Does not have to be a 50/50 split (situational sensitivity)

The So What:

- Fundamental key word "FIT"
- Think differently
- 5Cs from What does done look like?
- Achieve the right readiness
- The orchestrating role of the PMO
- Getting to a space where the learning is shared to find the right degree of hybrid implementation

Review Questions

Parentheses ( ) are used for Multiple Choice, when one answer is correct. Brackets [ ] are used for Multiple Answer, when many answers are correct.

Which of the following has been the most challenging for PMI to address in the matrixed way of delivering program work?

( ) Closing the gap between development and implementation
( ) Educating the functional organizations
( ) Certifying project managers
( ) Spreading the word across chapters.

Which of the following belongs to the 5C's of What does done look like?

( ) Complexity
( ) Candor
( ) Connective tissue
( ) Clarity

How can organizational experimentation become more successful?

( ) Stick to the leader's perspective of the program
( ) Centralize decision-making
( ) Try to add review gates
( ) Target a strategic split that enhances collaboration

CHAPTER 9

Risk-Based Governance

This chapter addresses the strategic elevation of governance practices, built on the foundation of having a balanced risk approach to enhanced program decisions. Risk management is a core strategic muscle for organizations. Many organizations remain unable to formulate risk properly or have it as a central stage for their governance and decisions. This is a typical lagging area of maturity that requires continual attention from the executive team and a consistent enablement for program stakeholders in order to better practice the proper risk management required skills.

Key Learnings

- Understand the skills necessary to design a risk-based program governance
- Learn how to gauge the risk appetite and practice cascading the risk management strategy across portfolios of programs and projects
- Explore the linkages between risk-based governance and achieving meaningful and faster decisions
- Understand the important tie between risk management and reinforcing the learning and maturing of the practices of program management in the organization
- Develop a strong awareness of the attributes necessary to make PMOs successful in achieving a mindset shift to governance that encompasses strategic agility

9.1 WHY RISK-BASED PROGRAM GOVERNANCE MATTERS?

Risk management matters. Risk-based governing is critical for managing programs life cycle and achieving timely and effective decisions. Addressing the resistance that is in the mind and practices of organizations to the topic and practice of risk management is critical. Program managers should drive getting the importance of this essential governance topic understood and socialized across their organizations' key stakeholders.

Program Management: Going Beyond Project Management to Enable Value-Driven Change, First Edition. Al Zeitoun.

Key Learnings

- The role of risk skills in leading and governing with excellence
- A Brené Brown example to clearly explain the importance of program accountability
- Why is risk-based governance the future of decision-making?
- Introducing the PMI-RMP as a differentiating professional certification for proactive governance (Becoming a Risk Management Professional)

9.1.1 Risk Skills for Leading and Governing with Excellence

Risk management is one of these knowledge and practice areas that show the true value of program/project management. It is about the proactivity and the look ahead into the future elements of the life cycle and become better equipped and prepared. As indicated in the PMI Standard for Risk Management, there are many topics of importance in the practice of risk management strategically. Among these key topics are addressing the risk attitude, tolerances, integrating with the program plans, and clear accountabilities for the process and for response strategics. In addition, like in any other process, there is a need for continuous enhancements.

Why risk is key to governance excellence?

- PMI addresses the importance of having open dialogues about risk early in the program (appetite and attitude toward risks)
- Risk ownership = governance excellence (as in the examples of working on or supporting a board – the fish stinks from the head and thus leaders have to be careful, the higher they go to allow for risk dialogues)

9.1.2 Brené Brown and Program Accountability

- Accountability is a vulnerable process (as in the video where she shows coffee spilling on her sweater and then blaming her partner who is not even there – showing also the need for vulnerability and not shifting blame)
- Program leaders own the process and the connectedness of team (we have the ability to make choices)
- This ownership allows for clear and effective choices (looking at the upside and downside and considering trade-offs)

9.1.3 Risk-based Governance and the Future of Decisions

What does decision-making look like in the future organizations and program teams?

- Fast
- Dynamic
- Often
- Consistent
- Across all layers of the organization
- Risk-centered (success in governance across industries is when the teams and teams of teams are open, use registers or checklists, check-ins, and multitude of ways to create ongoing interactions that take impediments out of the way)

FIGURE 9.1 The Collaborative Risk-based Future Decisions
Source: stux / 7312 images.

> **TIP**
>
> Program managers in the future are faced with a fast-paced, decision-making process and the expectation on the quality and nature of decisions need risk maturity.

9.1.4 PMI-Risk Management Professional (RMP)

- This risk management certification process expands you, and you become more holistic and able to look across the entire organization
- Creates a proactive mindset (fundamental shift)
- Enhances the understanding of the culture needed for program success
- Moves the practitioner and leader to a holistic understanding of linkages of uncertainty and opportunity to the strategic choices central to the digital future
- A clear professional differentiator (commitment to this certification shows a lot about the ability of the team to predict and look ahead)

The So What:

- Get excited about the PMI-RMP
- Risk-based governance is about having the right dialogues
- Proactivity is a key takeaway
- Taking fear out of the mix and encouraging conflicts discussions

Review Questions

Parentheses ( ) are used for Multiple Choice, when one answer is correct. Brackets [ ] are used for Multiple Answer, when many answers are correct.

_____ illustrates a set of the key skills that PMI address in the risk management standard.

( ) The program leader can delegate this topic to auditors
( ) Understanding the risk attitude and tolerance aids in proper risk management planning
( ) Governance is a board role only
( ) Team members should avoid risk dialogues

Which of the following are critical elements for proper understanding of accountability in program governance?
Choose all that apply

[ ] Only in the hands of the program manager
[ ] Continues to be a reactionary process
[ ] Program managers should ensure program process allows for it
[ ] Being a vulnerable process

Which is of the following is a clear sign of the future of decision-making?

( ) All decisions are left to program teams
( ) Slow decision-making process will prevail
( ) Decisions becoming highly risk-centered
( ) Executive makes the program decisions

9.2 THE CASCADING EFFECT OF THE RISK APPETITE

The critical element of the success of any practices is that they get cascaded across the organization. This could enhance the cross-organizational ability toward practicing risk-based governance. It is essential for program managers to exemplify these practices and to showcase the importance of cascading, and how it is done.

Key Learnings

- The best scenario – the right program manager's risk appetite strengthens governance excellence (requires investment and focus)
- Putting together the list of key factors that contribute to understanding risk appetite
- What are the factors that contribute to developing and cascading the risk appetite across programs and program teams?
- Showcasing the alignment in the execution of program work across teams using the UAE Nuclear Enterprise Program Management Office (EPMO) example of risk-based governance

9.2.1 The Right Program Manager's Risk Appetite!

The right program manager should have a balanced risk appetite. This will also help in creating the critical balance between running the business and growing the business. There is a need for sustaining success of operations, while stretching other practices and operating model changes that could make the program management environment more successful. The right mindset should be able to handle the program's VUCA environment and guiding the key risk management dialogues (Volatility, Uncertainty, Complexity, and Ambiguity are not going away).

As shown in Figure 9.2, it is important for the program managers to understand both sides of the coin for risk: threats and opportunities. In addition, they should work on knowing what their thresholds are for the program, the organization, and the stakeholders. This is about knowing how far we can go without escalation or having to count on expanded ownership for risks. This has to be coupled with the right attitude that encourages and invites the program team members to take on the necessary mistakes when needed, experiment as required, and learn fast for continual flow of program outcomes and benefits.

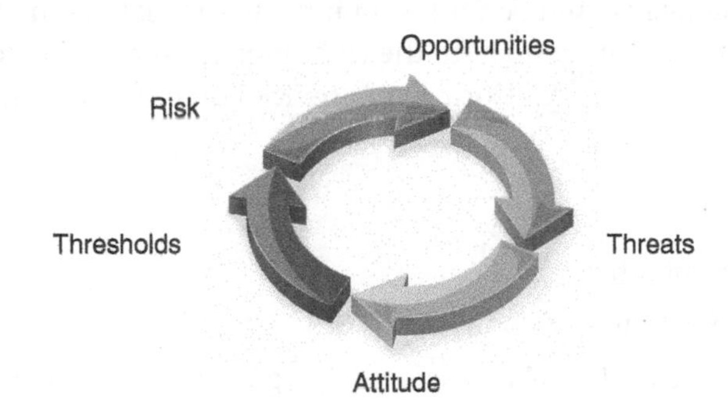

FIGURE 9.2 The Right Risk Appetite

Reminders

- Risk is front and center
- Encourage the right dialogue
- Attitude, not your aptitude, determines your altitude
- What are my boundaries without raising major flags?
- Look at these parameters as you set the risk appetite

9.2.2 Key Factors to Growing the Risk Appetite

- Turning a reactive culture into a proactive one begins with a commitment to owing continuous improvement of the risk management practices (entire life cycle of managing risks from identifying to analyzing to responding and continuously improving)

- Digital enablers and data analytics support the building of a risk appetite
- Stronger risk appetite creates speed and agility
- Strategic clarity, proper risk management policy, and a dynamic frame work supports the right appetite
- Going through a transformation requires facing complex decisions and associated transformation programs' risks (ensuring the right risk appetite links directly to transformation success)

9.2.3 Developing and Cascading the Risk Appetite

As highlighted in the PMI Standard for Risk Management, cascading of risks happens from the strategy to the portfolios, and then from the programs down to the projects. This connects the organization around practice of risk management across programs and provides a holistic view of ***Enterprise Risks***. In addition, this process has a feedback loop from the project up to the strategy where opportunities for enhancements and becoming more effective are highlighted and shared across the levels of strategy execution components.

- Portfolio into programs and projects (classic example of cascading)
- Enterprise Risk Management (ERM) requires the holistic view of risk and cascading our strategy
- Why is cascading critical for program managers? (Need to put the right value on a clear cascade)
  - Decisions autonomy
  - Connectedness to strategy
  - Cultural readiness

9.2.4 Showcasing Alignment in Execution of Program Work

As we review the dimensions of the UAE complex energy program success story previously highlighted, and as shown in Figure 9.3, program mangers could benefit from how these blocks come together to create a joint focus on program success. Purpose allows coming up with the right metrics and all the trade-offs needed to ensure the right value investment that drives the right platform for enhanced quality decisions.

The So What:

- Cascade is a key point to remember
- The connection between portfolios components
- Remember examples, such as in the UAE case, to help you develop a holistic view to govern with risk at the center

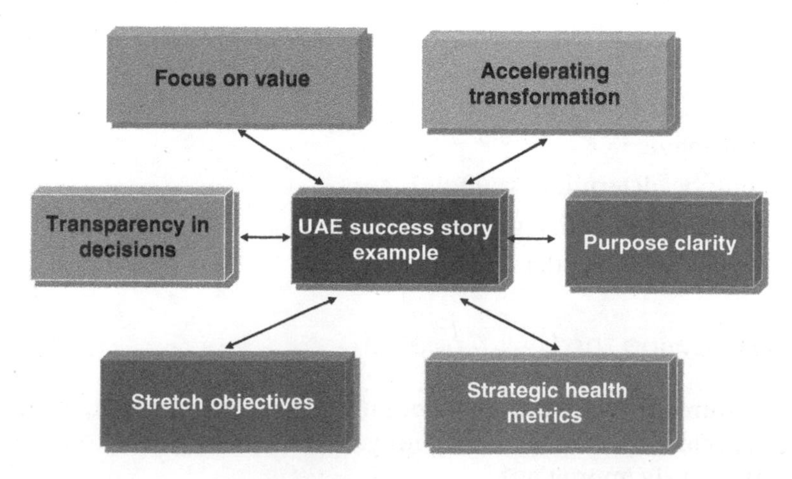

FIGURE 9.3 The UAE Success Story

Review Questions

Parentheses ( ) are used for Multiple Choice, when one answer is correct. Brackets [ ] are used for Multiple Answer, when many answers are correct.

_____ contributes to developing the right program manger's risk appetite.

( ) Head in the sand
( ) Understating the balance of threats and opportunities
( ) Unclear strategy
( ) A culture of denial

Which of the following most contributes to developing the risk appetite?
Choose all that apply

[ ] By enforcing the use of a risk register
[ ] Strategic clarity
[ ] By escalating all risks to management
[ ] An ERM framework

Which of the following is an example of how the right risk appetite contributes to successful execution?

( ) Increasing thc gates use in program life cycle
( ) Focus mainly on applying risk policy
( ) Supporting transparency in decision-making
( ) Using manual dashboards

9.3 DECISION-MAKING SPEED

Risk-based governance and speed have to go hand-in-hand. This might slow you down for a moment, yet it is the proactivity that you need most to ensure plans are thorough and create the most sustained outcomes. Speed of decisions is a great context to support risk-based governance. The PMO should be in the center of providing the leadership for this. Due to nature of risk practices and the human tendency to resist the active participation in this process, the PMO had to design and drive this as a change management process. PMO has to shift organizationally to get the focus on risk in the mix.

Key Learnings

- Creating alignment, re-informing the governance metrics around speed and quality of decision-making
- The postpandemic world's shift to PMOs that have their mindset focused on governance that encompasses strategic agility (fuel speed without sacrifices)
- Practical examples to illustrate how risk-based governance focus leads to meaningful and faster program decisions
- PMOs are adding product management to their capability

9.3.1 Governance Metrics and Speed of Decisions

- Governance metrics and speed are not opposites
- Investing the time to select the right metrics (while we go at speed)
- Future of work is value-central (is this helping?)
- Investing in and protecting critical conversations (Invest and protect those)
- Success is in creating alignment and operating with a **balance** of speed and quality (not about sacrificing – growth mindset orientation, e.g. Simon Sinek and John Maxwell and their views on this)

9.3.2 PMOs Mindset Shift to Strategic Agility

- Strategic agility is key to the work of the PMO and the practices across programs
- Executives are concerned with
  - The high % of failing programs
  - The amount of decisions escalation
  - Alignment of programs to strategy
  - Program work seen as tactical
- Shift to strategic agility
  - PMOs of the future are less about control or hand holding and mostly about driving and energizing strategic dialogues (adjusting strategy quickly and showing impact on roadmaps)

9.3.3 Risk-based Governance Focus

- A risk-based governance example:
  - Set strategy guardrails (enough linkages to governance and achieving strategy)
  - Decisions cascading magical formula = ***clarity of direction*** + ***autonomy*** (fit) + ***support for experimenting***
- Role of program manager is to remove any obstacles that would slow down the decisions (not slowing things down yet asking the right questions that make us more effective with the risks that we capture and handle early)

> **TIP**
>
> Program managers should practice the decisions cascading formula. Each of the formula's ingredients is core to the successful program culture.

9.3.4 Product Management Capability

Products are critical parts of the outcomes of many of the most strategic programs. For the PMO to have its full impact, it should be able to develop the product management muscles across program and product teams. As the coaching house, as highlighted in Figure 9.4, the PMO can combine the digital fluency capacity and the strong outcomes laser focus to achieve getting program teams to flow and to the anticipated speed of delivery:

- Continuous release for speed
- Product owners mindset in the PMO and thus the coaching house
- The more data analytics are used helps the PMO create a faster shift toward enhanced decision-making process

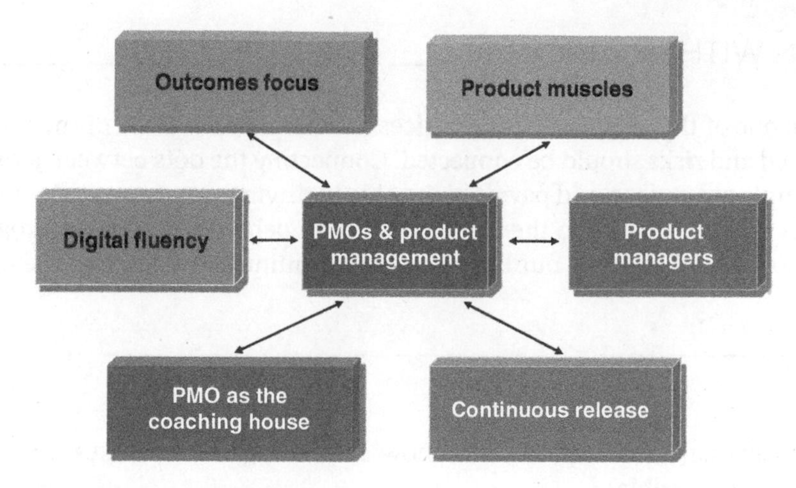

FIGURE 9.4 Role of PMO in Driving Product Management

The So What:

- Simple learning principles with the PMO focusing on mindset shifts such as product management
- Driving strategic agility using tools such as strategy guardrail
- Metrics should be focused on enhancing decisions speed without sacrificing what matters
- Risk and value go hand-in-hand

Review Questions

Parentheses ( ) are used for Multiple Choice, when one answer is correct. Brackets [ ] are used for Multiple Answer, when many answers are correct.

_____ illustrates that the governance metric choice will contribute to better program decisions.

( ) Keeping only executives in mind
( ) Investing in design of metrics based on achievement of value
( ) Focusing on increased control
( ) Minimizing the amount of dialogues

Which of the following concerns of executives drove the PMOs to shift their agendas to strategic agility? Choose all that apply

[ ] Gaps in aligning to strategy
[ ] Increased need of constraints control
[ ] The need to invest more in tactical solutions
[ ] the high percentage of escalation instances

Which is of the following could best reflect the focus of PMOs on building product management capabilities?

( ) Program managers are no longer needed
( ) The need for the PMO to operate as a coaching house
( ) Products are more important than projects
( ) Stakeholders demand it

9.4 INTEGRATION WITH LEARNING

Lessons learned remain one of the most valuable practices to build into the DNA of the program teams and their practices. Lessons learned and risks should be connected. Connecting the dots between a risk register and lessons learned register, as a simple example, could payoff high value in driving your impact as a program manager. Integrating across processes and tools is key to the value proposition behind your role as program manager. Setting our appetite around learning and building our learning agility continues to be one of the most admired future of work capabilities.

Key Learnings

- Levers that program managers can use to learn how controlling risk becomes moot if market demand shifts fluidly (Adapting is possible)
- How does the modern PMO's focus on outcome realization across business lines support strategic and learning agility?
- Illustrating how a risk-based governance approach can substantially enhance learning and delivering programs on time and on budget with the product or service needed (creating the future learning organization)

9.4.1 Market's Fluid Shifts

It is critical to understand the program's external environmental conditions and how they interplay in affecting the program's milestones and the balancing of risks that is necessary:

- Market shifts be very critical in their impact
- Agility as a critical future program manager muscle (being humble)
- Ownership of the dynamic nature of program risks (fluid)
- Growth mindset to weather the market storms (all the fluctuations and unrealistic expectations)
- Building the learning program team (every program team should have ownership of enhancing the learning for the organization)

9.4.2 Learning Agility

- Operating as a ***Learning Leader*** (a great title to have in order to connect the dots for the learning)
- Comprehending the program stories (understanding the context behind them)
- Mapping the ***horizontal workings around learning*** (new ways of working)
- Ownership of learning sharing and utilization (enough incentives and motivation needed for the program teams to do this)

TIP

Invest in becoming a *Learning Leader*.

9.4.3 Risk, Learning, and Delivering Successful Programs

- Myth behind risk's correlation to learning (connect the dots to be more open about enhancing risk management)
- Role of consistent learning from risk repositories (accessible, looked at, and utilized)

- Integrating risk in all release milestones (to create higher consistency and increase the dialogue around risks)
- Culture of continually improving delivery (having this mindset gets you a success formula that is not a controlling exercise, but more of a learning exercise)

The So What:

- Remember the importance of learning agility
- The linkages between risk and learning are critical
- Connection between this and changing the behaviors of the teams and the discussion amongst executives
- When learning prevails, this helps us tremendously in breaking down the siloed-mindset

Review Questions

Parentheses ( ) are used for Multiple Choice, when one answer is correct. Brackets [ ] are used for Multiple Answer, when many answers are correct.

_____ is an example of a lever that helps the program manager to make the most of risk management in a fluid market.

( ) Having rigid views
( ) Thinking positively
( ) Focusing on catching issues during execution
( ) Building a growth mindset

Which of the following are enablers for the PM to illustrate the value of learning agility? Choose all that apply

[ ] Getting executives to enforce learning
[ ] Building useful program stories
[ ] Breaking business segments silos
[ ] Being the boss

Which is of the following is a sign that risk and learning will contribute to program success?

( ) Increasing the investment in issue management
( ) Focus on empowering a risk committee
( ) Integrating learning to risk repositories
( ) Reviewing risks only at program closure

9.5 MATURING PROGRAM MANAGEMENT PRACTICE

Maturing is the way to recap this group of concepts related to this topic of risk-based governance. Finding ways for how to embed the ideas addressed above into the organization is what maturity consistently creates and where the PMO could shine. Success in repeating certain patterns or behaviors that matter should be part of the transformation roadmap crafter by the PMO and is core to the impact-driving mission of its members.

Key Learnings

- The PMO as the driver for program management maturity (who else could do it and connect the dots?)
- How to overcome the illusion of maturity is enough for program governance success?
- Identifying practical tips for creating repeatable healthy governance practices
- Analyzing PMI's pulse of the profession data around the 73% of programs and projects at organizations with high project management maturity meeting their original business goals and intent
- Reflecting on CMMI and other maturity models

9.5.1 PMO as the Driver for Program Management Maturity[1]

A natural role for the PMO to play and lead:

- The group of conductors are the PMs
- Owning the creation of consistent successful patterns
- Maturity as the path to strategic excellence (making sure that the PMO connects the dots)
- PMO could have the strategic roadmap for gradual movement to the maturity North Star

A key part of the driving impact of the PMO is to select the right methodology. There is a lot at stake in making these choices. Programs are generally considered as being composed of a series of projects. Historically, all of the projects within the program were executed by a one-size-fits-all methodology that often led to decision-making headaches for the program managers. However today, with the introduction of agile and Scrum, as well as other flexible methodologies, program managers have the option of selecting the best methodology available for each of the projects within the program. Each project part of a larger program may be executed using a different methodology. The difficulty facing the program managers will be in the selection of the methodologies to be used.

9.5.1.1 Understanding Methodologies

A methodology is a set of principles that a program manager can tailor and reduce to set of procedures and actions that can be applied to a specific situation or group of activities that have some degree of commonality. In a project or a program management environment, these principles might appear as a list of things to do and are often manifested in forms, guidelines, templates, and checklists. The principles may be structured to correspond to specific project/program life cycle phases, such as in a construction or a product development project.

The traditional project management or waterfall approach became the primary mechanism for the "command and control" of projects providing some degree of standardization in the execution of the work and control over the decision-making process. However, this standardization and control came at a price limiting those instances as to when this methodology could be used effectively. Typical limitations included:

- **Type of project:** Most methodologies that were either developed internally or purchased "off-the-shelf" assumed that the project's requirements were reasonably well-defined at the outset. As such, the project/program manager made trade-offs primarily based on time and cost rather than scope. This limited the use of the methodology to traditional or operational projects that were reasonably well-understood at the project approval stage and had a limited number of unknowns. Strategic projects, such as those that are part of program management activities involving innovation, where the end product, service or result, was much more difficult to define upfront, could not be easily managed using the waterfall approach because of the large number of unknowns and the fact that the requirements (i.e. scope) could change, and sometimes frequently.

[1] Part of this section has been adapted from a white paper, "Selecting the Appropriate Project Management Methodologies" by Harold Kerzner and J. LeRoy Ward.

- **Performance tracking:** With reasonable knowledge about the project's requirements, performance tracking was accomplished mainly using the triple constraints of time, cost, and scope. Nontraditional or strategic projects had significantly more constraints that required monitoring and therefore used other tracking systems than those offered by the methodology. Simply stated, the traditional methodology had extremely limited flexibility – and value – when applied to projects that were not operational.
- **Risk management:** Risk management is important on all types of projects. But on nontraditional or strategic projects, characterized by their high level of uncertainty and dynamic changes in requirements, many organizations found that the standard risk management practices included in traditional methodologies were insufficient for the type of risk assessment and mitigation practices found in such a fluid environment.
- **Governance:** For traditional projects, governance was often provided by a single person acting as the sponsor (if there even was one assigned!). The methodology became the sponsor's primary vehicle for command and control and used with the mistaken belief that all decisions could be made by monitoring just the project's time, cost, and scope constraints.

9.5.1.2 The Faulty Conclusion

Organizations reached the faulty conclusion that a single methodology, a one-size-fits-all approach, would satisfy the needs of almost all their projects. This mindset worked well in many companies where it was applied to primarily traditional or operational projects. But on nontraditional projects, the methodology failed, and in certain cases, in spectacular fashion.

Concurrent with the adoption and widespread use of the single methodology, strategic projects that included innovation, R&D, and entrepreneurship were being managed by functional managers who were often allowed to use their own approach for managing these projects rather than follow the one-size-fits-all methodology. Using innovation as an example, we know that there are several types of innovation projects each with different characteristics and requirements. Without employing a flexible or hybrid methodology, program managers were often at a loss as to the true status of these types of projects even though they were part of the program. Part of the problem was that professionals working on innovation projects wanted the "freedom to be creative as they see fit" and therefore did not want to be handcuffed by having to follow any form of rigid methodology or direction from the program manager.

9.5.1.3 Selecting the Right Framework

Deciding which approach or framework is best suited to a given project or program is a current challenge experienced by project/program managers in many, but not all, organizations. Some companies have not attempted to implement agile in any meaningful way and are still trying to solve all program issues with a "one-size-fits-all" approach. But the day is rapidly approaching where all program teams will be given the choice of which framework to use. We must never forget that the focus of our work on programs is on delivering value to our customer on a frequent basis. Whichever framework gets us there is the one we should be employing.

The decision regarding the best methodology requires answering several questions. Typical questions might include

1. **How clear are the requirements and the linkage to the strategic business objectives?** On certain programs, especially when innovation and/or R&D are required, it may be difficult to develop well-defined objectives even though the line-of-sight to the strategic business objectives is well known. These programs may focus more on Big, Hairy, Audacious Goals (BHAGs) rather than on more well-defined objectives.

   When the requirements are unclear or uncertain, the program may be tentative in nature and subject to cancellation. In short, it may look to many like an experiment where the team keeps pressing forward until certain events provide clear indicators that the projects within the program should continue or be terminated based on actual results. It is basically an exploratory endeavor. As such, we must expect that changes will occur throughout the life of the program. These types of programs require highly flexible frameworks and a high degree of customer involvement.

2. **How likely is it that changes in the requirements will take place over the life of the program?** The greater the expectation of changes, the greater the need for a highly flexible approach. Changes may occur because of changing consumer tastes, needs, or expectations. Allowing for too many changes to take place may get some of the project off track and result in a failed program that produces no benefits or business value. After all, even agile projects can suffer from scope creep. The size of the project/program is also important because larger projects/programs are more susceptible to scope changes.

   In addition to the number of changes that may be needed, it is also important to know how much time will be allowed for the changes to take place. In critical situations where the changes may have to implemented in days or weeks, a fast-paced, flexible approach may be necessary with continuous involvement by stakeholders and decision-makers.
3. **Will the customer expect all the features and functionality at the end of the project or will the customer allow for incremental scope changes?** Incremental scope changes allow the project to be broken down and completed in small increments that may increase the overall quality and tangible business value of the outcome. This may also provide less pressure on decision-making.
4. **Is the team colocated or virtual?** Projects that require a great deal of collaboration for decision-making may be more easily managed with a co-located team especially when a large amount of scope changes are expected.
5. **If the project requires the creation of features to a product, who determines which features are necessary?** The answer to this question may require the project/program team to interface frequently with marketing, clients, or end users to make sure that the features are what the users desire. The ease by which the team can interface with the end users may be of critical importance.
6. **Is there success (and/or failure) criteria that will help us determine when the project is done?** With poorly defined or an absolute lack of identified-success criteria, the project will most certainly require a great deal of flexibility, testing, and prototype development. In such cases, an iterative life cycle might be the best approach.
7. **How knowledgeable are the stakeholders with the framework selected?** If the stakeholders are unfamiliar with the framework, the program team may have to devote significant time to educate them on the framework selected and their expected role and responsibility in deploying that framework. Arguably, some might see this as a waste of time, but if the stakeholders are not sure of their roles and responsibilities, it makes the program team's job that much more difficult. Therefore, training and education on agile methods needs to be provided to all impacted in the organization, not just the facilitator, product owner, or development teams. It is hoped that providing training and indoctrination in the new frameworks will lessen the resistance so often found in people who cling to the old ways of doing things.
8. **What metrics will the stakeholders and business owner require?** Waterfall methodologies focus on time, cost, and scope metrics. Flexible methodologies also allow for other metrics such as business benefits and value achieved.

9.5.1.4 Be Careful what you Wish for

Selecting the right framework may seem like a relatively easy thing to do. However, all methodologies and frameworks come with disadvantages as well as advantages. Project teams must then "hope for the best" but "plan for the worst." They must understand what can go wrong and select an approach where execution issues can be readily resolved in a timely manner.

Some of the questions focusing on "What can go wrong?" that should be addressed before finalizing the approach to be taken include

1. Are the customer's expectations realistic?
2. Will the needs of the projects/program be evolving or known at the outset?
3. Can the required work be broken down and managed using small work packages and sprints or is it an all-or-nothing approach?

4. Will the customer and stakeholders provide the necessary support, and, in a timely manner?
5. Will the customer and stakeholders be overbearing and try to manage some of the project themselves?
6. How much documentation will be required?
7. Will the project/program team possess the necessary communications, teamwork, and innovation/technical skills?
8. Will the team members be able to commit the necessary time to the program?
9. Is the type of contract (i.e. fixed price, cost reimbursable, cost sharing, etc.) well suited for the framework selected for some of the outsourced projects?

Selecting a highly flexible approach may seem, at face value, to be the best way to go since mistakes and potential risks can be identified early allowing for faster corrective action thereby preventing disasters. But what many seem to fail to realize is that the greater the level of flexibility, more layers of management and supervision may need to be in place.

9.5.2 Illusion of Maturity Is Enough

There continues to be a vast number of maturity models to refer to and use around the globe. Organizations and individuals seeking maturity levels should always be on guard that is not yet another quality measuring approach to assess following processes or certain practices. It has no direct full guarantee that the outcomes of the program are going to be most effective. It is almost tike the previous work we covered under the definition of success and the rethinking of success from the classical world of projects and activities to the world or programs and value. Figure 9.5 highlights the typical building blocks that are seen across a majority of the effective models.

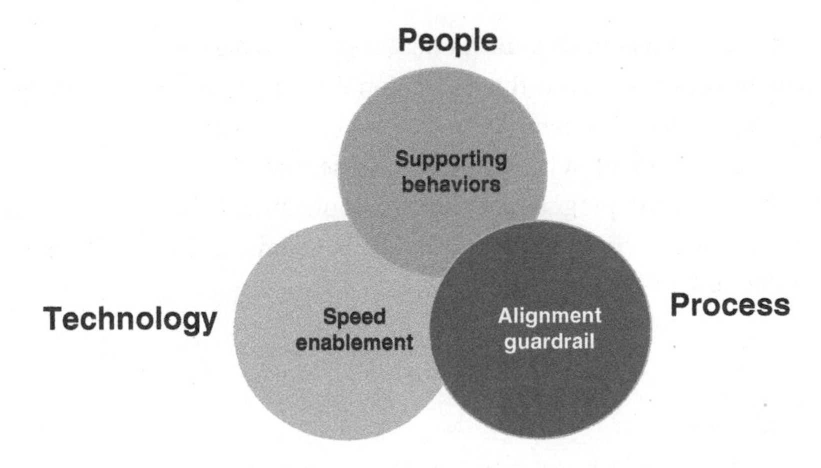

FIGURE 9.5 Building Blocks of Maturity Models

- Classic maturity models are usually ranking progress on a scale from 1 through 5
- Linking back to what matters, being the strategy
- Maturity success is in enhancing our ability to achieve outcomes faster

9.5.3 Creating Repeatable Healthy Governance Practices

- Higher maturity of PM practice enhances enterprise governance
  - Vulnerability becomes easier (at the leadership level and all the way to the teams)
  - Bouncing back on track is more likely
  - Govern with trust and empathy (no limitation across boundaries)
  - Role of leadership (with a servant mindset)
  - Value of consistency in governing with confidence (enough of the right supporting info)

9.5.4 Programs at Organizations with High Maturity

Source: geralt / 25607 images.

- PMI indicates that 73% of programs and projects at organizations with high maturity meet business goals
- Routines:
  - Discipline
  - Accountability (modeling that, e.g. what we discussed from Brene Brown)
  - Consistency of cross-organizational principles (repeatable fashion as in the risk-based governance approach)

9.5.5 Capability Maturity Model Integrated (CMMI) and other Models

- Also, Kerzner's and many others in Organizational Program Management
- The acknowledgement across models is the realization that they drive improvement focus, yet
  - Teams have to practice what they commit to
  - Truthful and sincere investment in improvement is essential
  - Research shows that executive program sponsorship contributes to higher value of maturity achievement
  - Tailoring models to organizational readiness (how ready am I and committing to the steps in order to make the implementation very practical)

The So What:

- Maturity is the main theme
- Create the foundation for the risk-based approach to governance
- Looks at people, processes, and technology
- No matter the model, key is the fit and readiness
- The right investment of time and effort to support the maturity
- PMO can help us create that consistency of implementing repeated patterns

Review Questions

Parentheses ( ) are used for Multiple Choice, when one answer is correct. Brackets [ ] are used for Multiple Answer, when many answers are correct.

_____ is an example of a way by which the PMO would lead the program maturity journey.

( ) Always ensuring everyone on program teams uses the same templates
( ) A North Star strategic roadmap
( ) Ensuring the strategy does not change
( ) Escalate maturity to the program sponsor

Which of the following contributes to building healthy governance practices?
Choose all that apply

[ ] Policing
[ ] Governing with trust
[ ] Consistent patterns in decision-making
[ ] Increasing the role of the leader in making decisions

Which is of the following is critical for a maturing model to drive future programs success?

( ) Focus on achieving more maturity certifications
( ) Start immediately regardless of readiness
( ) Investing in continual improvement
( ) Use the same model for all parts of the organization.

CHAPTER 10

The Learning Engine

This chapter addresses a strategic core capability of the PMO, namely learning. Future organizations are like ***learning engines***. With the changing way of doing business and the strategic role programs and projects play, executive leadership is finally realizing the golden opportunity it has to use their initiatives as a way to anchor the foundation of learning across lines of the business. The ***horizontal way of working*** these initiatives require and allow these initiatives' leaders to turn their roadmaps and plans into distinct milestones for learning.

This does not come easy as there are still remaining stigmas from the cultures of the older generation organizations, where some residual of command and control remains. The PMO has to step up and drive this learning transformation. Documenting, analyzing, sharing, and engaging are four of the focus areas PMOs have to master to create fluidity of the learning exchanges. In tomorrow's enterprises, the executives, and even the board of directors are the ones to set the tone for this learning engine and have to make it a strategic priority. The PMO has the ownership of executing that strategy and making it about value and not a checking the learning hours box exercise.

Key Learnings

- Understand the impact of creating a learning enterprise
- Get immersed into the changing role of the program offices as an End-2-End (E2E) enabler of capabilities, tools, and processes.
- Learn the direction that PMOs take to create a learning culture
- Explore a case where the power of data activates faster learning and supports delivery of solutions to market with speed and accuracy
- Develop the continuous improvement skills to enhance value flow focus and guide you to deliver greater program benefits

10.1 THE ENTERPRISE LEARNING MUSCLES

Future PMOs should make it a priority and a high focus area to become dedicated to learning and getting deeper into building and growing this learning muscle. There are natural reasons why the PMO is the right entity to lead this.

Program Management: Going Beyond Project Management to Enable Value-Driven Change, First Edition. Al Zeitoun.

This relates to the typical integrating role of the PMO, the number of initiatives it is engaged in, and the trust placed in the PMO by the organizational executive leaders. Just like building physical muscles, learning is a choice, and it also affects the health of the enterprise. Investing in this engine, opens the door to a growth mindset, where positive behaviors prevail and the capacity to turn dreams and ideas into reality happen. Thus change programs.

Key Learnings

- Why are the PMOs positioned to become the enterprise learning engines of the future?
- How does learning muscle support how organizations actively and holistically "design" their customer experience? (What does it take to develop that?)
- The role the PMO plays in operating as End-2-End (E2E) in terms of capabilities, tools, and processes.

10.1.1 Enterprise Learning Engines of the Future

- They drive the key strategic questions (most natural part of the PMO function)
- They have visibility of the entire enterprise (PMO has E-2-E view)
- They provide a coaching style in relating
- Operating in a sponge-like learning entity enhances PMO's chance for creating impact (mindset that the PMO uses daily)
- The greater the autonomy of the PMO, the more likely it becomes the connecting learning engine! (Cascading that mindset across the program teams)

10.1.2 Holistically "Design" the Customer Experience (CX)

- Learning muscle requires fluidity in adapting to most fitting framework (learning and driving future programs)
- CX is core to building the learning muscle with the customer central to every practice
- CX-centered learning prepares the enterprise to lead and deliver around value (nothing is as important in programs than ensuring that this is achieved)

10.1.3 End-to-End (E-2-E) Capabilities, Tools, and Processes

This is how the PMO plays its role well. With that E-2-E focus, the PMO sets a holistic foundation for learning. Figure 10.1 highlights the strategic learning lifecycle steps the PMO can use to partner and engage with the customers, program managers, program teams, and other key stakeholders to spread the learning mindset and succeed to cut across the enterprise:

- Data-driven assessment (rigor needed)
- Strategizing with core program teams (cocreation and inclusion of cross teams' ideas)
- Creative collaboration with partner
- Energizing with capabilities, tools, and processes (experimenting)
- Successful and timely strategy enhancements (This is how the PMO plays its strategic impact role on the organization)

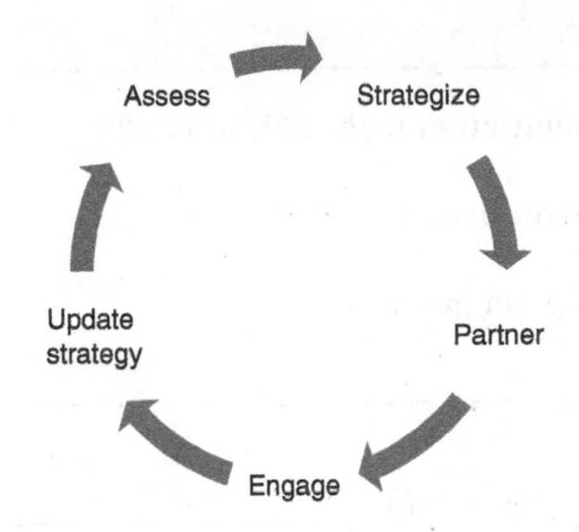

FIGURE 10.1 E-2-E Strategic Learning Cycle

> **TIP**
> The PMO has a great strategic opportunity to drive building a growth mindset by imbedding learning as integral part of the programs' way of working.

The So What:

- What is the meaning of a learning engine?
- The flexibility needed around frameworks
- The dynamic nature of learning with the experimentation and data use at the center
- The PMO has to drive by example with the right mindset and the idea of the lab we discussed a few times in previous lessons

Review Questions

Parentheses ( ) are used for Multiple Choice, when one answer is correct. Brackets [ ] are used for Multiple Answer, when many answers are correct.

_____ is an example of how the PMO becomes the learning engine.

( ) Asking a lot of questions
( ) Hosting the key strategic questions
( ) Program managers have their home in the PMO
( ) Using enforcing style in relating

Which of the following are signs that CX enhances the quality of the learning engine? Choose all that apply

[ ] Increased investment in doing whatever the customer wants
[ ] Leading around value
[ ] Drive the change in a direction chosen by the customer
[ ] Adapting to fitting frameworks

(continued)

(continued)

Which is of the following is a contributor to the E2E success?

( ) Sticking to one time program approach enhancement
( ) Being technology-centric
( ) Creatively collaborating with program partners
( ) Creating a large program team

10.2 DEVELOPING ROLE OF THE PMO

The PMO and the program managers community should work on turning the theory behind being the learning engine into reality. We recall the saying, tell me how I am measured, I will tell you how I will perform. This would mean that the PMO should work with management and the learning and development entities in the business to drive the cross-organizational understanding of this role, its value, and how the incorporation of learning targets are directly ties to team members and other professionals' growth journeys. Ultimately, the PMO is a collection of program managers and team members who have the purpose of successful use of program and project management practices in the center of their mission. It is key to develop the PMO to take on that strategic role.

Key Learnings

- Using simulation to capture learning and activate the learning engine and achieve delivery at speed and accuracy (connecting the dots to technology)
- Review the capabilities needed for the PMO to play a more evident strategic role around learning (as in the strategic dialogues discussion)
- The Siemens example of automated Product Lifecycle Management to illustrate the power of data (why this illustrates the power of data in achieving speed and higher quality decisions)

10.2.1 Simulation and Learning

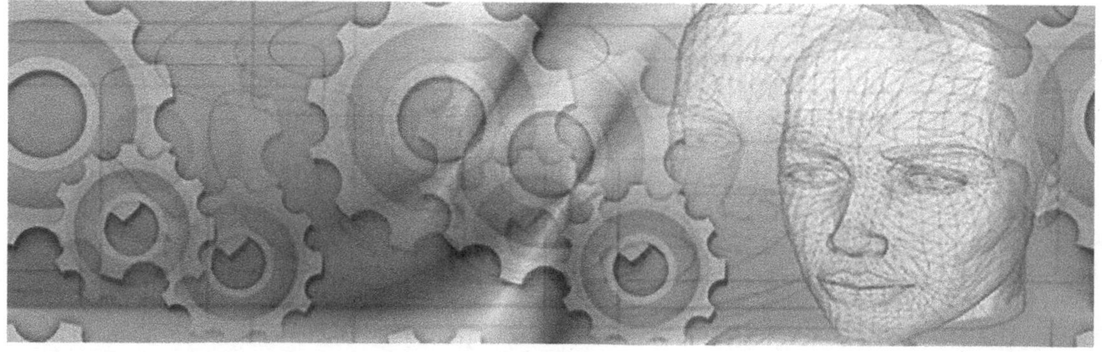

Source: kalhh / 4296 images.

- **Simulation**
  - Predict program and product performance (extracting useful analytics across complex data sets)
  - Fluidity of customization and choices (music to your ears as PMs)
- **Learning**
  - Fast decision-making (connection between what is real and what is virtual)
  - Trust to accelerate learning (simulation enables the PMOs to bring this learning as a strategic priority for the organization)

10.2.2 Strategic Role

Figure 10.2 brings to the forefront the very critical and valuable strategic role PMOs could play. An innovation lab is basically the organizational heart for experimenting, creating safe spaces, utilizing technology, and creating digital threads for the products and other supporting services the organization might be involved in. Of course, the basic foundational aspects for the success of this role have got to be in place. Trust, coupled with great insights, using design thinking, and strengthening the sensing and responding capabilities, could all come together as the right ingredients for this role.

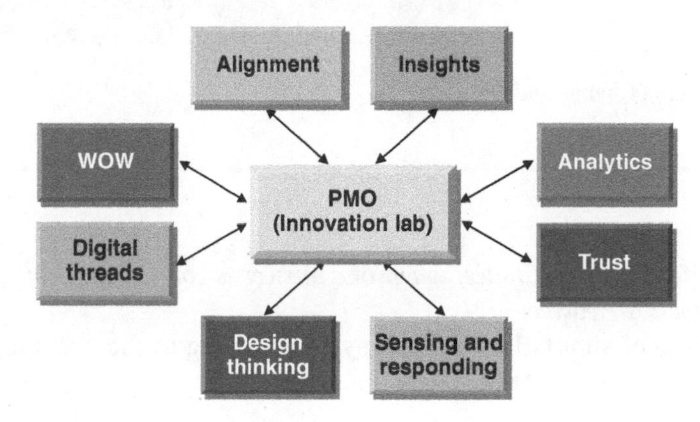

FIGURE 10.2 PMO as the Innovation Lab

- Affects the environment in order to being the learning organization
- Looking at how the entire integration of our solutions affects our ways of working
- Strategy is about choices
- Learning matters the most in order for the PMO to drive the innovation success of the organization

> **TIP**
>
> The PMO could drive organizational excellence by playing its strategic role as the future *Innovation Lab*, where programs' data and learnings affect impactful change.

10.2.3 The Power of Data

- Figure 10.3 is a good reminder of the balance needed for the powerful use of data: It is ultimately about people
- Need to make sure that digital in the mix allows me to adapt and customize better due to the power data brings as a critical part of my learning engine

The So What:

- The PMO should use the technology and data such as simulation to connect the dots
- Establish the link where the learning engine helps us cresting the strategic organization we aspire to be

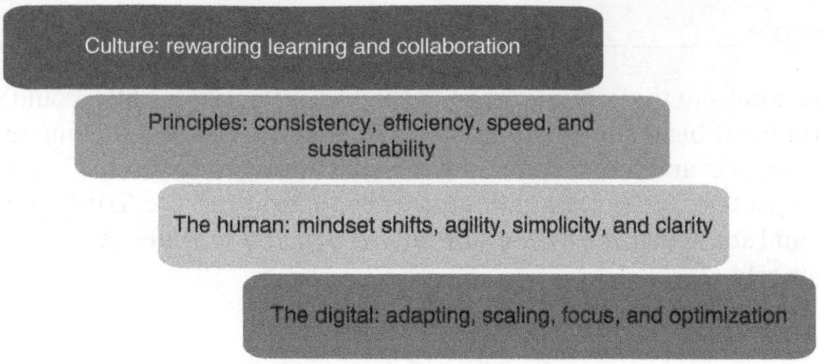

Culture: rewarding learning and collaboration

Principles: consistency, efficiency, speed, and sustainability

The human: mindset shifts, agility, simplicity, and clarity

The digital: adapting, scaling, focus, and optimization

FIGURE 10.3 PMO and Balancing Change and Data

Review Questions

Parentheses ( ) are used for Multiple Choice, when one answer is correct. Brackets [ ] are used for Multiple Answer, when many answers are correct.

_____ is a critical attribute of simulation technology contributing to the accuracy of delivering valuable solutions.

() Saving development time
() Fluidity in customizing products
() Cutting costs
() Focus on the engineering details

Innovation requires failure. What would the PMO focus on that would enhance its strategic role in driving innovation?

Choose all that apply

[ ] Provides stronger alignment
[ ] It enhances our WOW choices
[ ] Creates the safety needed for frequent experimentation
[ ] Effectiveness of firefighting

Which is of the following is an example of the power of data's impact on scaling outcomes?

() Higher focus on getting products to market faster
() Expanding the structured processes use
() A program team finds a new idea for adjacencies
() Only build what we are good at

10.3 CREATING THE LEARNING CULTURE

Peter Drucker's famous "Culture eats strategy for breakfast" remains a relevant reminder that most good intentions and strategic plans could fall apart when the organizational culture opposes the forward movement with these great ideas. PMOs for the future have to get the culture right. When it comes to learning, transparency and data enablers support the right culture shift we need to have. This culture maturing role is yet another valuable impact the PMO could drive.

Key Learnings

- Building the learning culture and transparency value
- Developing the understanding that the learning culture directly supports enhanced program execution capabilities
- How could the PMO to align its portfolio of work to strategies and learning, the business using a single PMIS to support the dissemination of well-analyzed learning? (Much more impactful to work across programs and workflows such as in eth case of organizations like ServiceNow and others)

10.3.1 Transparency Value

- Value of open strategic dialogue for future Program Management
- Role of building high performance program teams (e.g. the five layers of Lencioni and especially that second layer about handling conflicts)
- Transparency questions:
  - Are there ***better ways*** of looking at this?
  - What are we ***missing***? Great question to creating vulnerability
  - Are there biases? Allowing us to fix some of our ***wrong perceptions***

TIP

The PMO should practice the use of transparency questions to help in building the learning culture foundation.

10.3.2 Learning Culture and Execution

Learning Program Culture Features

- Servant leaders equipped with learning mindset enable effective planning around value
- Emphasis on benefits with stronger cross-functional collaboration creates strategic focus

Enhanced Program Execution Features

- Program requirements are fluid and thus execution outcomes are supported by continuous learning.
- Context-driven planning allows for expedited tailoring along the program life cycle.

10.3.3 Could Learning Limit Innovation?

- Could lead to short-term decisions affecting value (patience is required)
- Could result in negative behaviors if misused (need to out the energy in this)
- Could minimize the vulnerability required (otherwise we could lose sight of the innovation we want to create and thus the importance of EQ)

10.3.4 PMIS and Learning

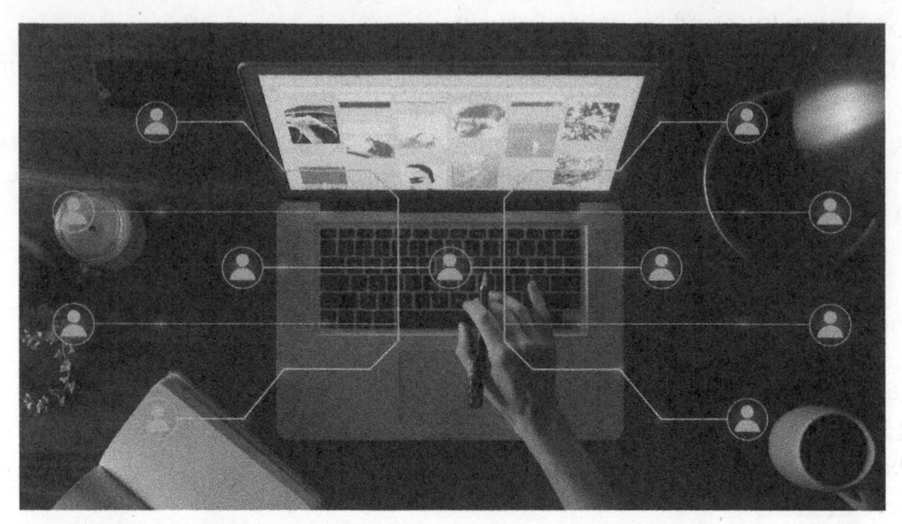

Source: Buffik / 9 images.

Establishing and maintaining a fitting Program Management Information System (PMIS) is a critical enabler for the success of the PMO in driving the execution of strategy.

- The importance of the infrastructure as a key investment
- Automation for faster insights and tilting
- Enhanced Business Experience (the entire experience including the PMs, the teams, the supply chain, and others)
- Unified source of truth for a companywide view (very powerful if we can get aligned views)
- Resources optimization as we move to project economy (critical for productivity and the shift that is taking place to the changing the business side of the equation)

The So What:

- Building transparency and driving the right question
- Develop you own list of questions
- The patience that is required or the long-term investment thinking
- A PMIS is a key investment in our programs' success

Review Questions

Parentheses ( ) are used for Multiple Choice, when one answer is correct. Brackets [ ] are used for Multiple Answer, when many answers are correct.

_____ illustrates the importance of the beginner's mind in tackling complex program challenges.

( ) Higher focus on tactical program work
( ) Asking powerful questions
( ) Ease of digital use
( ) Reaching consensus fast

Which of the following is an illustration of the learning culture's value to enhanced execution? Choose all that apply

[ ] Eliminating biases
[ ] Context-driven planning
[ ] Best behaving program teams
[ ] Cross-functional collaboration opportunities

Which is of the following is a direct advantage of investing in a PMIS?

( ) Defend our program approach
( ) Unified source of truth for a companywide view
( ) High focus on audit functions
( ) Clearly established business rules

10.4 CRITICALITY OF CROSS-PROGRAMS ALIGNMENT

Program management secret sauce remains all about investing in and succeeding in utilizing a matrixed-way of working. The future PMO owns this changed way of working implementation.

Key Learnings

- Using the powerful questions to ensure the key learning is extracted (very important in your role)
- Understanding the PMO role in enterprise-level program planning
- The importance of cross-alignment as businesses continues to move closer to a "project economy"
- Building cross-programs understanding of the value of learning (key muscles to develop)

10.4.1 Powerful Questions

The PMO could develop a series of lists of questions to address various contexts and drive different aspects of organizational change. A sample of these powerful questions could include

- What is the level of our diversity and inclusion of different and opposing ideas as necessary? (The more diverse the better and providing equity to different voices – natural aspect of PM's role)
- Are the insights, themes, direction implications limiting us? Are we creating enough distance as needed? (To minimize biases)
- How are we ensuring that a growth mindset is prevalent and our ability to scale is not limited? (Fundamental question)

10.4.2 Enterprise-Level Planning

- Way of Working (WOW) should be anchored in learning
- As highlighted by figure 10.4, program use of context-driven planning and value linkages is an important muscle for the PMO to develop

Consistency
Value
WoW

FIGURE 10.4 PMO and Progress toward Consistency

- Consistency across the program teams:
  - Strategy
  - Principles
  - Source of truth (as in the case of useful data and insights)

The Project Economy

- Need for a new WOW (where learning becomes a natural)
- Continual learning drives limitless transformation (every project or program serves transformation or a change and that is where learning is feasible)

Value of Learning

- When program stakeholders have a joint dream, they learn how to inspire teams for action
- Learning provides courage for important conversations which would create authentic harmony across program teams (mindset of alignment around value and optimization of resources usage)

The So What:

- Cross collaboration across the program teams is a must
- Start with the proper questions
- Inspire teams around the true North Star
- If we connect the dots and focus on value with the true transparency, we have a chance to uncover cross organization WOW that is needed to tackle complex programs

Review Questions

Parentheses ( ) are used for Multiple Choice, when one answer is correct. Brackets [ ] are used for Multiple Answer, when many answers are correct.

_____ illustrates the three building blocks of sustained program performance.

( ) Consistency, repeatability, and control measures
( ) WoW, value focus, and consistency
( ) start, stop, and continue
( ) Value, strong alignment, and communications

Which of the following are key ingredients of the Human 2.0 equation? Choose all that apply

[ ] Toughness in program dialogues
[ ] Sensing the changing needs of program stakeholders
[ ] Ideating without limits
[ ] Maintaining program teams' energy and passion

Which is value of the practice of stop, start, and continue?

( ) Strengthening the program team's structure
( ) Focus mainly on catching failures
( ) Creating a balance of focus on learning
( ) Disregarding time pressures

10.5 GUIDED CONTINUOUS IMPROVEMENT

Great topic to showcase the PMO's impact. Continuous improvement has always been an important part of any process and in the case of program and project processes, that is not an exception. The additional word here "Guided" is an important addition to show the role of being customized, coached, and delivered in a way that could fit and enhance learning and drive change. The focus on value, mindset shift to risk, and the learning engine are all part of the PMOs strategic future role. When the PMO guides the improvement, it becomes much more fitting and impactful for the organization success. Programs are dealing with different changes, and thus guiding what fits where they are very helpful.

Key Learnings

- Using the learning capability to source the skills needed to fulfill emerging program needs
- The role of the PMO in prioritizing areas of improvement and encourage Communities of Practice (COP) utilization (The Spotify example)
- How would value flow focus guide where continuous improvement will matter most? – a key of this lesson
- Using stories to complete the stop, start, continue learning worksheets (fundamental to the changes in the learning behaviors)

10.5.1 Emerging Program Needs

Source: WOKANDAPIX / 1264 images.

- Resilience is what helps us (as when we addressed this topic previously)
- VUCA environment continues to drive changing needs
- Massive amount of data that learning allows us to understand and put to good use, coupled with the power of AI and ChatGPT types of growing technologies
- Move to strategic dialogue and value focus enhances PMO's responsiveness (need to move the needle on this)
- For example, a customer leadership change (PMO can help a lot in this change chaos)
  - program manager is able to connect with useful insights
  - demonstrate programs' value
  - achieve effective transitions (PMs supported by the PMOs bring stability to the mix coupled with the growth and use of EQ)

10.5.2 Prioritizing Improvements

Reminders

- As in the case of a retrospective
- Openness to various frameworks
- Continual learning appetite
- Spotify and COP living (teams and teams of teams learning and creating practices that can spread as a living nature)
- Use various techniques: Must-have, should-have, could-have, and won't have (MoSCoW), paired comparison (different inputs of priorities across groups of stakeholders)
- Practicing dynamic priority shifts with the E-2-E view (what matters in the success of a portfolio of programs and projects)

10.5.3 Practicing Guided Continuous Improvement (GCI)

- Have guardrails around strategy that are adjustable
- Value dialogue ensures that we are not limited to pure process focus
- Figure 10.5 is an additional reminder that all continuous improvement efforts need guardrails
- Needs to be guided to strategically important areas

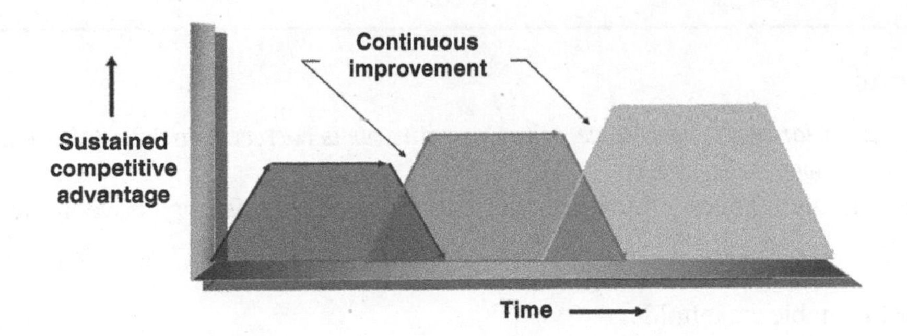

FIGURE 10.5 GCI

10.5.4 Stories and Associated Start, Stop, Continue

- As referenced often, figure 10.6, shows a very powerful tool for program managers
- Stories connect us to context fast (e.g. the 5Cs: Color, Context, Connective Tissue, Cost, and Consequence)
- Are the insights from stories helpful in creating a path forward strategy adjustment?
- How do we create consensus on investing, divesting, or get more details? (Or more research exhibiting the qualities of a dynamic learning team)

The So What:

- The importance of guiding the continuous improvement
- The PMO brings back the strategic dialogues to the center
- The use of guardrails
- Prioritization techniques to help us focus on what matters the most
- This is mainly where the PMO could become the learning engine of the future bring back trends and practices to the mix
- Enhancing the strategic dialogue which makes the PMO the most successful strategy execution arm for the organization

Stop	Start	Continue Learning

FIGURE 10.6 Start-Stop-Continue Tool

Review Questions

Parentheses ( ) are used for Multiple Choice, when one answer is correct. Brackets [ ] are used for Multiple Answer, when many answers are correct.

______ is a technique that enhances our prioritization driven by distinguishing between the must do and the nice to do.

( ) Opinion of most valuable stakeholder
( ) MoSCoW
( ) Paired comparison
( ) Weighting system

Which of the following helps continuous improvement investments become guided? Choose all that apply

[ ] Listen to the views of management
[ ] Bringing the dialogues to value
[ ] Providing a guardrail
[ ] Electing to implement many improvement ideas

Which is of the following is an example of the value of stories in making better decisions?

( ) Making the program team feel better
( ) Coming up with our own assumptions
( ) Reaching consensus on investing, divesting
( ) Mastery of learning

SECTION IV

ORGANIZATIONAL CHANGE MANAGEMENT FRAMEWORK – TRANSFORMING STRATEGY EXECUTION TO REALIZE PROGRAM VALUE

SECTION OVERVIEW

This section addresses one of the most talked about topics in business today, change. As you know everyone wants change and growth, yet the discipline, skills, and tactics necessary are often missing. The lessons and ideas in this chapter will expedite your learning and ability to be on a clear path for building alliances and roadmaps, and become more empowered to drive the most complex changes in your work and programs environments.

You will get the opportunity to learn and understand the healthy change culture attributes and be ready to operate as a leader for change and what it takes to sustain change outcomes. You will learn a multitude of practices that strengthen your strategy execution muscles. These insights will encourage you to create new fresh views of success that strategically change future business outcomes.

SECTION LEARNINGS

- The growth path for program metrics and their use in creating a mature focus on what change matters most
- Examples and case studies that get you to deeply understand the importance of driving change anchored in ownership, and the future role of change scientists
- Understand the skills behind your critical differentiator future role as a leader of organizational change
- Next-generation work practices fluency
- Critical tools addition to your impact toolbox

Program Management: Going Beyond Project Management to Enable Value-Driven Change, First Edition. Al Zeitoun.

KEY WORDS

- Change Culture
- Ownership
- Change Scientists
- Adapting Factor
- Co-creation

CHAPTER 11

Change Culture

This chapter is focused on explaining the change management culture and the related attributes for consistency in governance. The world has changed and hybrid work has become the dominant normal. This has direct implications on program work and program teams. As leaders and program team members, it is crucial to understand the cultural changes surrounding this new normal and find the most fitting ways to excel across the organization and program teams.

Key Learnings

Understand the change culture and its criticality for programs success.

- Get introduced to the attributes for healthy governance and enterprise rick management (ERM).
- Explore an organizational case that exemplifies sources of transformation success commitment.
- Learn the host of ERM practices that are key to the ways of decision-making across the organization.
- Address the blockers to benefits relation in programs.

11.1 THE FEATURES OF CHANGE CULTURE

Change cultures thrive on change. They welcome change and create a muscle of attracting and supporting the achievement of change outcomes. These cultures are dynamic and adaptable. They also represent many of tomorrow's organization cultures, where the norm is fluidity of expectations, requirements, and customers' increasing demands.

Key Learnings

- What does it take to create the value and impact in the change culture?
- Enablers, strategies, and tools to help us create impact across teams
- Actionable series of change topics
- Builds on Peter Drucker's statement: Culture east strategy for breakfast

Program Management: Going Beyond Project Management to Enable Value-Driven Change, First Edition. Al Zeitoun.

11.1.1 Value-Driven Way of Working

Highlights PMI's Benefits Realization Management (BRM)

- BRM discipline sphere of influence is large!
  - What does it take to create a change culture?
  - The standard for the practice of this important value to programs and their management
  - Dedicated thinking to the entire lifecycle
  - Be very clear on what success looks like
  - The importance to taking the time upfront
  - Looking ahead to what it takes to sustain change outcomes from a given program (an area we typically fall short at)
  - Like the example of a manufacturing plant when I have rolled out new technologies yet fell short in getting the right ownership
- Multiple players to create impact: an army of stakeholders are required:
  - Executive Leaders (critical commitment element)
  - Sponsors (they drive the movement to value)
  - Benefit Owners (certain experiences preparation, practices, and expectations over a maturity journey) – delegation is required to that person to own the achievement of benefits
  - Change Agents (they create and communicate the necessary engagement culture)
  - Program Leaders (responsible for connecting the dots for change)
  - Many more – depending on the nature of the program and its complexity

11.1.2 Program Transformation Example

- Building accountability and the Benefits Realization Muscles
  - Achieving the change across transformation examples is built on clarity of expectations
- Excellence in Leadership
  - They got to realize that we will have to make some tough calls and decisions to do what is necessary to stay the course across the change lifecycle
- Clarity of Communications for Quality Decisions
  - Right processes and leadership are enabled with the right technologies that enhance the ongoing focus on sharing the transformation progress and impact

11.1.3 BRM Critical Success Enablers (CSEs)

What does it take to be successful?

- Roles and responsibilities (we all deal with this even in the success of our programs and projects)
- Culture (values, communications, and joint commitment to the change's benefits)
- Skills (what questions to ask)
- Adaptability (right ownership will support this and allow us to shift and adjust to ensure we are on a relevant journey – common sense, yet not highly practiced)
- Risk-based governance (fundamental to the value of programs success, shows you are living in the reality of the programs and you are planning the right readiness)
- Traceability (think about the criticality of how a given change/benefit is mapped to the right reason that drove the journey of a given change program)

The So What:

- Having clarity of ownership and dialogues and realizing the lifecycle focus requires a strategic set of steps to drive change and its success
- Need to raise the hand and ask for the support when gaps are being faced

Review Questions

Parentheses ( ) are used for Multiple Choice, when one answer is correct. Brackets [ ] are used for Multiple Answer, when many answers are correct.

_____ takes place throughout the benefits realization lifecycle.

( ) Program management outputs delivered.
( ) Portfolio management outputs and optimization.
( ) Identify benefits.
( ) Sustain benefits.

What contributes the most to the success of creating a change management culture? Choose all that apply

[ ] Centralized leadership.
[ ] Intentional communications.
[ ] Focus on quality metrics.
[ ] Building the right muscles.

Which is of the following is not an example of a benefits realization critical success enabler?

( ) Issues-based governance.
( ) Skills.
( ) Adaptability.
( ) Culture.

11.2 CHANGE SUCCESS INGREDIENTS

When we talk culture, we have to believe it is feasible to change it.

The worst scenario – no culture support for driving changes necessary for executing strategy.

Key Learnings

- Why culture support for BRM matters?
  - The required dedication to programs' value by multiple stakeholders (ownership by creating buy-in across groups of stakeholders)
  - The shift in nature of continual value dialogue (a key priority that the PM needs to keep an eye on)
  - The role of executives and value owners across the organization (the tone has to be set at the top and execs should always ask for the necessary change commitment)

11.2.1 Focus: Process, Leadership, Awareness, and Communication

- Processes to manage lifecycle and secure alignment (simplification and revisiting our processes and their value)
- Leaders' commitment to BRM (are the actions following what they are committing to and are the rights skills being supported)
- New awareness of benefits responsibility (e.g. creating a customer experience requires joint clarity and commitment coupled with change management skills)
- Manage resistance with Organizational Change Management (OCM) skills (the listening, the stepping back, and collaboration across program teams)

11.2.2 Strategic New Ways of Measuring Value

- Getting to consistency around value (shifting from the traditional time, cost, and quality to what is important to the business, like the financials, and other concrete growth outcomes)
- Intangibles are critical in the new ways of working: collaboration, EQ, morale, sustainability, quality of life, etc. (important for the value to stick, e.g. in a manufacturing plant besides the direct deliverables, the satisfaction of clients and employees is what sets us on a path towards using the strategic indicators)

TIP

Measuring value is truly a journey that requires a conscious focus on creating a way of working that will help us sustain the achievement of benefits.

11.2.3 Program Management Discipline and Achieving Benefits

This can be applicable in any delivery framework space:

- Seeing customers value in different light
- Clear delegation to benefits owners to make tough calls to course correct (speed is critical)
- Move to different business model and organizational structure might be necessary (tough calls and the clear roles and responsibilities ones in play here)
- Horizontal ways of working are necessary (programs and achievement of change hinges on our ability to stick as on organization to this)
    - Shift in mindset (ongoing commitment)
    - New motivation models (work with learning and development)
    - Rethinking processes to move towards strategic metrics (as in the model shown in Figure 11.1 and requires a top engagement of the stakeholders)

The metrics maturity model can be used as a practice exercise to look at what we are practicing. If we are at level 1 or 2, what is it going to take to move up the steps? Work with colleagues and others to achieve that. Mindset shift and processes rethink is an ongoing practice.

The So What:

- Change culture tone is set at the top
- Rethink our views of metrics and how to progress across the metrics maturity journey

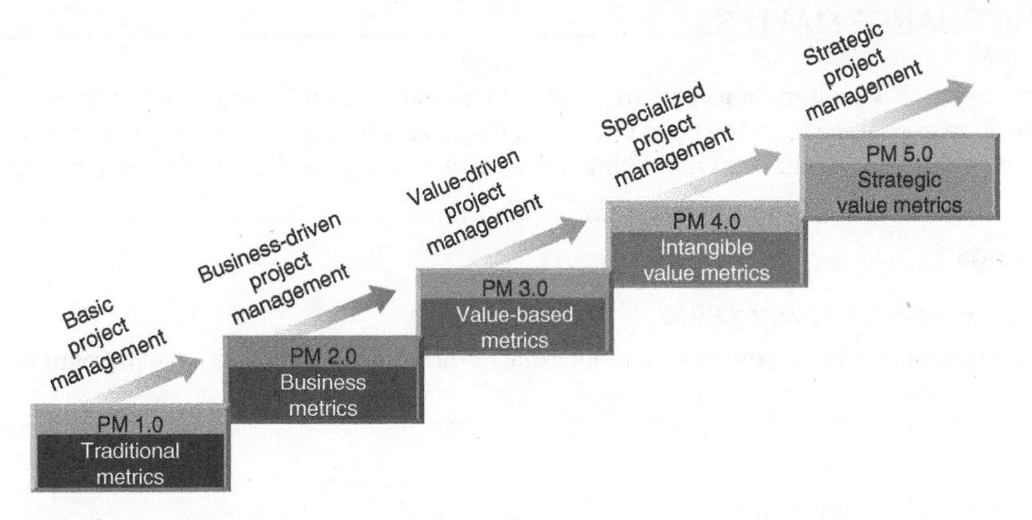

FIGURE 11.1 Metrics Maturity Journey
Source: Adapted from Kerzner, Harold, Zeitoun, Al and Viana Vargas, Ricardo *Project Management Next Generation: The Pillars for Organizational Excellence*. Wiley, 2022. p. 440

Review Questions

Parentheses ( ) are used for Multiple Choice, when one answer is correct. Brackets [ ] are used for Multiple Answer, when many answers are correct.

_____ illustrates that an organizational culture is supportive of focus on programs benefits.

( ) The increasing interest in the triple constraints.
( ) Time for sufficient ongoing benefits dialogue.
( ) Discussing benefits only at the onset of a program.
( ) Team members continue to be rewarded on deliverables completion.

Which of the following are direct contributors to achieving the cultural focus on change outcomes success?

Choose all that apply

[ ] Awareness of benefits responsibility in PM's role.
[ ] Build processes on tighter governance.
[ ] Give resistance time to dissolve.
[ ] Leaders' commitment to BRM.

Which is an example of an intangible metric?

( ) Earned Value.
( ) Sales volume.
( ) Stress level.
( ) User onboarding time.

11.3 GOVERNANCE MATTERS

Sometimes governance is thought of as policing. It should be about the elements in our culture that will help us move the needle to ensure we have the right outcomes. It's about finding the right level of program governance that fits the needs of the program team in making the necessary change and the realization of benefits happen.

Key Learnings

- Suitable governance for a healthy BRM
- The orchestra conductor as a powerful reminder that at the end of the day we have the right leadership in connecting the dots

11.3.1 Healthy BRM Muscles

When the organization builds this muscle, the focus is on a fitting governance:

- Central value orchestrating and focus role (orchestrator view)
- Enablers for reaching consensus (when empowered the change leaders helps us create that focus)
- Closing the gap between deliverables and value (team members could be delivering their pieces, yet it's about how the collective piece of music reaches the ears of the audience)
- Ensuring clarity of roles and responsibilities for value (each role has to be in harmony)
- Addressing hybrid delivery models (what works the best and thus the fit is what matters and what has to be adjusted to meet the needs for the change in a healthy culture)

11.3.2 Excellence in Governance

- Continuous improvement effort – decision-makers have to make it a priority
- Time is spent on significant issues that show value matters (very critical)
- Decision-makers have an adequate opportunity to participate in critical discussions
- Diversity of expertise and perspectives is assured (key for the right dialogues)
- Different views are listened to in a constructive and courteous manner (it is not about silence or passive aggressive views but about positive effective dialogues)

11.3.3 Right People in the Room and Effective ERM

Having the right people in the room when it comes to achievement of program benefits can't be emphasized enough!

- Rethink the stakeholders (who are the right ones in the mix)
- Ensure time prioritization on value (in the work of the program and the discussions around the work of the program)
- Cancel non-value add programs (also applies to benefits)
- Ensure executives' leadership of ERM (executive agenda is centering on ERM and having risk central to the mix)
- Develop decision-making intelligence (right people equipped with the right technology, AI, Power BI, and anything that helps with trends and threads the make decisions more effective, thus achieving the right excellence in governance)

The So What:

- Governance does matter – mostly because when we get to excel in it and have the right degree of it, that will help us adapt to the changing needs of the stakeholders
- The orchestra view helps us ensure all the team players are focusing on value and that we have right people on the room to enable the right decision-making and prioritization

Review Questions

Parentheses ( ) are used for Multiple Choice, when one answer is correct. Brackets [ ] are used for Multiple Answer, when many answers are correct.

Which of the following is a sign of a balanced governance?

( ) Closing the gap between opposing views.
( ) Educating customers on governance structure.
( ) One size fit all.
( ) Spreading the work to highest achievers.

Which of the following belongs to excellence governance behaviors?

( ) Socializing time.
( ) Faster decisions.
( ) Diverse views recognition.
( ) Allow every decision-maker as much time as needed.

How can organizational ERM become more successful?

( ) Stick to quick consensus around responses.
( ) Centralize decision-making.
( ) Add more review gates.
( ) Ensure executives' leadership of ERM.

11.4 ERM-BASED GOVERNANCE

Why is risk so important to governance and has to be tackled so many times for it to stick and be put in practice. This is such a core for project and program management's success, given the nature of the uncertainty associated with initiatives seeking change creation.

Key Learnings

- Being pragmatic
- Data-driven
- Consistent in our processes
- The practical ways by which we have a higher chance for achieving change success

11.4.1 BRM Framework Practical Stages

- *As reflected in figure 11.2, BRM starts with Identify*: Critical to the success of this framework is the early diverse dialogues with leadership
- *Execute*: The work of the portfolio follows (achieving and reprioritizing)
- *Sustain*: The ongoing interaction with the stakeholders who are keen on realized benefits (the ones who are keen on achieving the Initially identified program benefits)

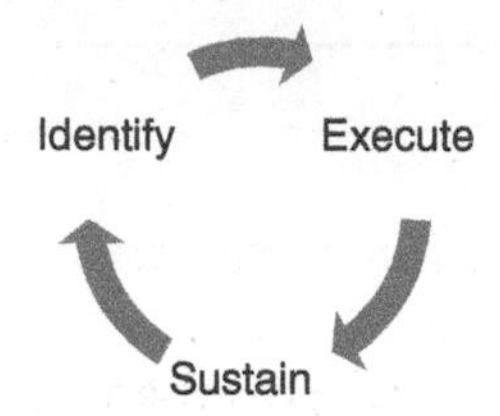

FIGURE 11.2 BRM Stages

11.4.2 The Program Manager with an ERM Role

- Understand how to work the culture and build on organizational ways of working that has ERM as a strategic muscle
- Use digital to support decisions and create time to think (the importance of capitalizing on digitization to create more effectiveness)
- The right risk appetite to ensure speed and agility (ways to adapt along the way given the time that was created)
- Strategic clarity (what matters at the end of the day and how the program contributes to achieving the strategy)
- Get coaching/mentoring support as relevant (understanding how to look at the program end-to-end)

11.4.3 The Healthy BRM Culture and the Proper Governance

Healthy BRM is built on the collaboration and cross connection like what we need in program teams, even in rainy and tough times.

- ERM requires the holistic view of strategy (bigger picture and the opportunity on the other side)
- What matters in BRM culture?
  - Better business case (protecting its achievement)
  - Proper governance level (asking the right questions)
  - Cultural proactivity and awareness (this is why risk focus is in the mix and that we discuss what matters early on)

11.4.4 The Benefits Profile

Figure 11.3 shows a good way to look at what builds the profile based on ongoing engaging, the importance of clarity early and going back to ask the tough questions, talking strategically to address what could hinder our success early, looking at surprises across components and business units, and building on the learnings form the metrics maturity steps and the trust and digital enablers we have addressed.

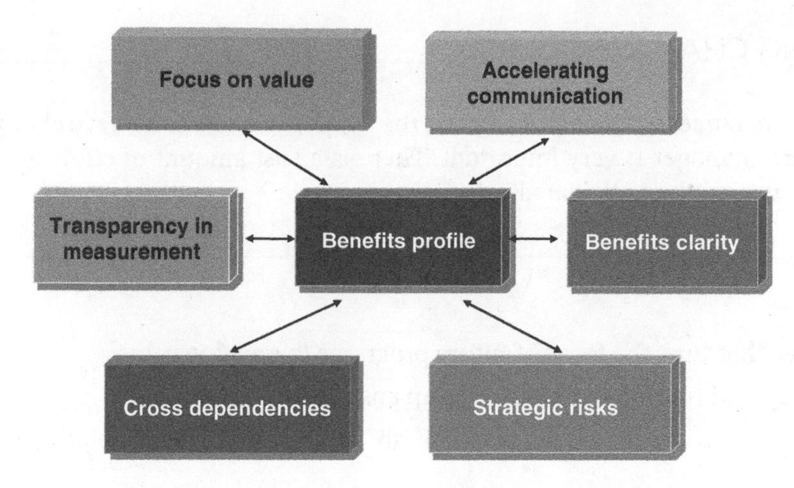

FIGURE 11.3 Benefits Profile

The So What:

- The enterprise view of risk is fundamental to governance
- Program manager should make sure that the benefits profile helps bring the important elements together
- Having a DNA in our way of working that demonstrates investing time correctly and uses digital to aid where we spend our time, is critical

Review Questions

Parentheses ( ) are used for Multiple Choice, when one answer is correct. Brackets [ ] are used for Multiple Answer, when many answers are correct.

_____ is the benefits framework stage that might require multiple iterations.

( ) Execute.
( ) Sustain.
( ) Identify.
( ) None of the above.

Which of the following directly contributes to developing the program manager's ERM capabilities? Choose all that apply

[ ] By spending more time with the team.
[ ] Strategic clarity.
[ ] By escalating all risks faster.
[ ] Use digital to support decisions.

Which of the following is an example of how the benefits profile ensures ERM focus?

( ) Increasing the number of metrics used.
( ) Focus mainly on applying risk policy.
( ) Capturing strategic risks and cross dependencies.
( ) Using multiple dashboards.

11.5 SUPPORTING CHANGE SUCCESS

Many stakeholders have to come together and align on the support needed to achieve change success. This is why this role for the program manager is very important. There is a vast amount of effort to be invested in aligning around the change's focus and the anticipated befits.

Key Learnings

- Choose the practices that turn the focus of future programs to creating value
- How do benefits map and traceability matrices help ensuring BRM success?
- Use a BRM Readiness Survey to self-asses where an organization stands relative to critical factors for BRM success
- Understand the root causes and dismantling the blockers for benefits realization success across the categories of stakeholders, planning, processes, and team

11.5.1 Turn Program's Focus to Value

As in putting a house purchase together. Not an easy exercise just as in the case or real estate and hoping to achieve the dream of the right home ownership.

- Owning the change and clarity of expectations (just like in the case of home buying experience)
- Establishing accountability and diversity of roles and responsibilities (all the players such as legal, finance, others that have to come together)
- PMOs shift to strategic results and impact creation (got to be impact creators focused on the achievement of strategic value metrics as discussed and using the works "so what" and the questioning mindset that thinks about why are we doing it!)

11.5.2 Role of Benefits Maps and Traceability Matrix

A key part of achieving change is to visualize the change and how the different pieces of the puzzle have to connect to create the ultimate beautiful picture. A benefits map shows the nice flow between outputs to the final impact. On should take the time and use the right tools like mind maps to connect the dots between what you do and achieving value. Additionally, the traceability builds on using benefits register and looks across different programs to see which ones are used to achieve what benefits, including possible primary and secondary focus of select programs. Visuals are great to create the ownership needed.

11.5.3 BRM Readiness

In the nuclear industry example, there was an entire organization that ensures that the regulators' needs are addressed, training is done, and operational steps are all properly checked. Readiness is a big thing. The program team thinks it is ready and it might not be!

Diverse inputs are essential to ensure cross organizational support for critical success enablers (CSEs)

- Framework and culture are setup well (simplicity)
- Fitting variety of skills (diversity is critical)
- Govern with agility (in addition to ERM, know when I need to adapt and shift)
- Role of risk management
- Sufficient benefits tracking (what are the mechanisms such as benefits register, automated tools, etc.)

11.5.4 Dismantling Benefits Blockers

- PMI indicates in its pulse of the profession studies that 73% of programs and projects at organizations with high maturity meet business goals
- Dismantling blockers:
  - Change Focus Discipline (making sure that there is the right focus on value)
  - Executive Leadership (where are they spending their time)
  - Consistency of Benefits Thinking (a must have)

The So What:

- The traceability matrix, making sure we have clarity of the entire benefits journey
- Stopping and taking the time to discuss and challenge the team, and creating a committed dedication to taking obstacles out of the way
- Culture eats strategy for breakfast

Review Questions

Parentheses ( ) are used for Multiple Choice, when one answer is correct. Brackets [ ] are used for Multiple Answer, when many answers are correct.

_____ is an example of a way by which benefits maps contribute to enabling benefits success.

( ) There is no real use, just more paperwork.
( ) Aligning outputs to outcomes then to benefits and goals.
( ) Ensuring the strategy is fixed.
( ) Escalate topics continually to the program sponsor.

Which of the following could be captured in surveys to support change achievement success? Choose all that apply

[ ] Policy setup.
[ ] Governing approach.
[ ] Consistent tracking.
[ ] Increasing the role of risk managers.

Which is of the following is most critical for dismantling blockers of future programs success?

( ) Focus on risk management certifications.
( ) Start immediately regardless of readiness.
( ) Investing in executive leadership commitment.
( ) Use the same framework for every program.

CHAPTER 12

Sustaining Benefits

This chapter is focused on sustaining change outcomes and the role of the program manager in creating the change lifecycle resilience. Benefits realization lifecycle is dependent on the nature of the program, and thus the anticipated journey to realize these benefits could vary. Having a clear ownership for sustaining benefits and the right supporting plan is a must for this to happen. Multiple stakeholders have to come together and align on their roles for what will be required for sustaining benefits and agreeing on what that would look like.

Key Learnings

- Understand the lifecycle view of identifying, executing, and sustaining benefits.
- Reflect on examples for creating the resiliency needed for managing the programs benefits.
- Learn how to create the environment, the ownership, and the trust for sustaining benefits.
- Explore the HBR study on reasons how strategy execution falters.
- Develop an appreciation for how a global consultancy' secret sauce for cascading ownership could work for you and your organization.

12.1 BENEFITS ACROSS THE LIFECYCLE[1]

12.1.1 Introduction

Organizations have been struggling with the creation of a portfolio of projects within a program that would provide sustainable business value. All too often, companies would add all project requests to the program queue for delivery without proper evaluation and with little regard if the new projects were aligned with program business objectives or provided benefits and value upon successful completion. Projects were often submitted without any accompanying business case or alignment to business strategy. Many projects had accompanying business cases

[1]Material in this section provided by Harold Kerzner

Program Management: Going Beyond Project Management to Enable Value-Driven Change, First Edition. Al Zeitoun.

that were based upon highly exaggerated expectations and unrealistic benefits. Other projects were created because of the whims of management and the order in which the projects were completed was based upon the rank or title of the requestor. Simply because an executive says "Get It Done" does not mean it will happen. The result was often project failure, a waste of precious resources, and in some cases, business value was eroded or destroyed rather than created.

12.1.2 Understanding the Terminology

Before continuing on, it is important to understand the basic terminology.

A ***benefit*** is an outcome from actions, behaviors, products, or services that are important or advantageous to specific individuals, such as customers, business owners, or a group of individuals, such as stakeholders. Benefits might include:

- Improvements in quality, productivity, or efficiency
- Cost avoidance or cost reduction
- Increase in revenue generation
- Improvements in customer relations and customer service
- Maintaining a firm's customer base
- A means for finding new customers

Not all program benefits are directly related to program profitability but may be influenced by the sustainability needs for the program. The benefits, whether they are strategic or nonstrategic, are normally aligned to the program's business objectives of the sponsoring organization that will eventually receive the benefits. The benefits appear through the ***deliverables or outputs*** that are created by the projects. It is the responsibility of the program manager to participate in the selection of the right projects and then assign project managers to create the deliverables.

Benefits are identified in the project's business case. Some benefits are tangible and can be quantified. Other benefits, such as an improvement in employee morale, may be difficult to measure and therefore treated as intangible benefits.

There can also be dependencies between the benefits where one benefit is dependent on the outcome of another. As an example, a desired improvement in revenue generation may be dependent upon an improvement in quality.

Benefits realization management is a collection of processes, principles, and deliverables to effectively manage the organization's investments.[2] Both program and project management focus on maintaining the established baselines, whereas benefits realization management analyzes the relationship that the program has to the business objectives by monitoring for potential waste, acceptable levels of resources, risk, cost, quality, and time as it relates to the desired benefits.

Decision-makers must understand that, over the lifecycle of a program, circumstances can change requiring modification of the requirements, shifting of priorities and redefinition of the desired outcomes. It is entirely possible that the benefits can change to a point where the outcome provides detrimental results, and the program should be cancelled or backlogged for consideration later. Some of the factors that can induce changes in the benefits and resulting value include:

- *Changes in business owner or executive leadership:* Over the life of a program, there can be a change in corporate leadership or program leadership. Executives that originally crafted the program may have passed it along others that have a tough time understanding the benefits, are unwilling to provide the same level of commitment, or see other programs as providing more important benefits.

[2] For additional information on benefits realization management see Letavec (2004) and Melton et al. (2008)

- *Changes in assumptions:* Based upon the length of the program, the assumptions can and most likely will change, especially those related to enterprise environmental factors and customer needs and expectations. Tracking metrics must be established to make sure that the original or changing assumptions are still aligned with the expected benefits.
- *Changes in constraints:* Changes in market conditions (i.e. markets served and consumer behavior) or risks can induce changes in the constraints. Companies may approve scope changes to take advantage of additional opportunities or reduce funding based upon cash flow restrictions. Metrics must also track for changes in the constraints.
- *Changes in resource availability:* The availability or loss of resources with the necessary critical skills is always an issue and can impact benefits if a breakthrough in technology is needed to achieve the benefits or to find a better technical approach with less risk.

Program ***value*** is what the benefits are worth to someone. Programs generally have a longer duration than projects and determination of a program's value may require years of measurement. Project or program business value can be quantified, whereas benefits are usually explained qualitatively. When we say that the ROI should improve, we are discussing benefits. But when we say that the ROI should improve by 20%, we are discussing value. Progress toward value generation is easier to measure than benefits realization, especially during project and program execution. Benefits and value are generally inseparable; it is difficult to discuss one without the other.

12.1.3 The Business Case

Benefits realization and value management begins with the preparation of the business case. There are four major players in benefits realization and value management activities:

- A governance committee composed of members that possess at least a cursory level of knowledge of both program and project management
- The benefits or business owner
- The change management owner if organizational change management is necessary to harvest the benefits
- Project and/or program managers

The business owner, possibly accompanied by the program manager, is responsible for the preparation of the business case as well as contributing to the benefits realization plan. Typical steps that are included as part of business case development are:

- Identification of opportunities such as improved efficiencies, effectiveness, waste reduction, cost savings, new business, etc.
- Benefits defined in both business and financial terms
- A benefits realization plan
- Estimated project costs
- Recommended metrics for tracking benefits and value
- Risk management
- Resource requirements
- High-level schedules and milestones
- Degree of project complexity
- Assumptions and constraints
- Technology requirements; new or existing
- Exit strategies if the project must be terminated

Templates can be established for most of the items in the business case. A template for a benefits realization plan might include the following:

- A description of the benefits
- Identification of each benefit as tangible or intangible
- Identification of the recipient of each benefit
- How the benefits will be realized
- How the benefits will be measured
- The realization date for each benefit
- The handover activities to another group that may be responsible for converting the project's deliverables into benefits realization

12.1.4 Measuring Benefits and Value

The growth in metric measurement techniques has made it possible to measure just about anything. This includes benefits and value. But at present, since many of the measurement techniques for newer metrics are in the infancy stage, there is still difficulty in obtaining accurate results. Performance results will be reported both quantitatively and qualitatively. There is also difficulty in deciding when to perform the measurements, incrementally as the projects progress or at completion. Measurements on benefits and value are more difficult to determine incrementally as the projects progress than at the end.

Value is generally quantifiable and easier to measure than benefits. On some programs, the value of the benefits of the program cannot be quantified until several months after the program has been completed. As an example, a government agency enlarges a road to hopefully reduce traffic congestion. The value of the project may not be known until several months after the construction project has been completed and traffic flow measurements have been made. Value measurements at the end of a program, or shortly thereafter, are generally more accurate than ongoing value measurements as the program progresses.

Benefits realization and business value do not come from simply having talented resources or superior capabilities. Rather, they come from how the organization uses the resources. Sometimes, even projects with well thought out plans and superior talent do not end up creating business value and can even destroy existing value. An example might be a technical prima donna that views this project or program as his/her chance for glory and tries to exceed the requirements to a point where the schedule slips and business opportunities are missed. This occurs when team members believe that personal objectives are more important than business objectives.

Once program deliverables are created, benefits harvesting occurs, which is the actual realization of the benefits and accompanying value. Harvesting may necessitate the implementation of an organizational change management plan that may remove people from their comfort zone. Full benefit realization may face resistance from managers, workers, customers, suppliers, and partners. There may be an inherent fear that change will be accompanied by loss of promotion prospects, less authority and responsibility, and possible loss of respect from peers.

Benefits harvesting may also increase the benefits realization costs because of:

- Hiring and training new recruits
- Changing the roles of existing personnel and providing training
- Relocating existing personnel
- Providing additional or new management support
- Updating computer systems
- Purchasing new software
- Creating new policies and procedures
- Renegotiating union contracts
- Developing new relationships with suppliers, distributors, partners, and joint ventures

Program managers desire to have their program last as long as possible and be profitable. However, there must be a tradeoff when sustainment of the program may require changes that disrupt the ongoing work. Expanding a product line with new products or adding new features to existing products may require major changes to the manufacturing facilities and possibly result in having to close out old facilities. Workers may be laid off or must undergo training in how to work differently. Simply stated, there are risks with wanting to extend the life of a program.

12.1.5 Causes of Complete or Partial Failure

No matter how hard we try to become good at benefits realization and value management, there are always things that can go wrong and lead us to disaster. Fourteen such causes of failure that can occur along the entire lifecycle of a program might include:

- No active involvement by the business owner or stakeholders
- Decision-makers are unsure about their roles and responsibilities, especially in the early lifecycle phases
- Projects are approved as part of a program but without a supporting business case or benefits realization plan
- A high level of uncertainly and ambiguity exists in defining the benefits and value such that they cannot be described adequately in a document such as a benefits realization plan
- Highly optimistic or often unrealistic estimates of benefits are made to get project approval and a high priority
- Failing to recognize the importance of effective resource management practices and the link to benefits realization management
- Maintaining a heavy focus on the program's deliverables rather than on benefit realization and the creation of business value
- Using the wrong definition of program success
- Failing to track benefits and value over the complete lifecycle
- Not having criteria established for when to cancel some failing activities within the program
- Having no transformational process, if necessary, where the benefits and value can be achieved only from organizational change management
- Failing to capture lessons learned and best practices, thus allowing mistakes to be repeated

Because of the importance of benefits and value, today's program managers are more of business managers than the pure project or program managers of the past. Today's program managers are expected to make business decisions as well as program-based decisions. Program managers today seem to know more about the business than their predecessors.

12.1.6 Benefits Realization Cube

The lifecycle approach to benefits builds on the elements of the culture and the enablement that we put in place for achieving the change benefits. It is critical to sustain the benefits beyond the lifecycle of point of delivering the outcomes of a given program. The people, process, and other enablers and efforts required have to all come together in a clear integrated plan for this to succeed.

There are multiple ways to categorize the benefits that the program manager and program stakeholders need to sustain. This could be three-dimensional as discussed in the following text.

Source: qimono / 504 images.

- Tangible or Intangible benefits have to be in balance (importance of intangibles to measuring value given the small amount of attention given to it)
- Planned or Emergent (emerging from the right dialogues, like putting a fan at the end of a manufacturing line to figure out which package has not been properly filled out)
- Direct or Indirect
- A Cube is a great way to visualize how a combination of these benefits types (multi-dimensional and thus the set of combinations affect the design for achieving benefits and the possible outcomes)

12.1.7 BRM Across Program Lifecycle

Program stakeholders could have different views of benefits from the inception and along the program lifecycle phases. Program managers will need to manage that and follow an adaptable approach to deal with the uncertainty, updates, and the impacts of continual innovation, as highlighted in Figure 12.1. This adaptability contributes to strengthening the innovation muscle!

> **TIP**
>
> Ask the right questions, use different ideas, replan along the way, and practice the creative and innovative mindset to get back on track for benefits achievement

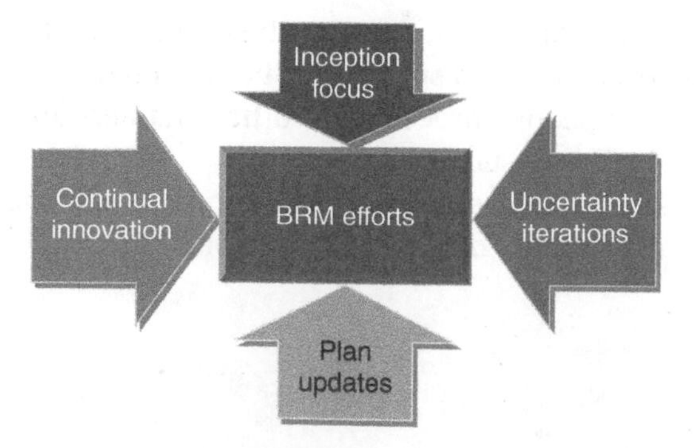

FIGURE 12.1 BRM Efforts

12.1.8 Benefits Register's Value

Regardless of the lifecycle delivery approach, or the changes in benefits along the program's journey, the program manager has a tremendous opportunity to go back and revisit the elements of the register and its use.

Benefits Register Elements

- Description
- Categorization
- Time horizons
- Ownership
- Metrics

Considerations at the ***Portfolio Level*** (looking at things strategically so the register is giving you the focus, visibility, and connecting the dots to the benefits traceability and helping the benefits owners ensure achieving and sustaining benefits).

The So What:

- The cube principle affects and shapes benefits expectations
- The importance of the replanning and rethinking muscle
- The discipline around the use of adaptive and continuously updated benefits register to sustain the benefits beyond the completion of the program

Review Questions

Parentheses ( ) are used for Multiple Choice, when one answer is correct. Brackets [ ] are used for Multiple Answer, when many answers are correct.

_____ is an example of the direct value of using a benefits cube in your program work.

( ) Which benefit should the team report on?
( ) Visualizing how an overlap of benefits tiers affects prioritization.
( ) Pictures are powerful.
( ) How much funding in a program benefit should be made?

Which of the following contributes the most to the effectiveness of BRM efforts across the program lifecycle? Choose all that apply

[ ] Maintain one BRM plan throughout the lifecycle.
[ ] Early focus on formulating benefits measurements.
[ ] Showing off exceeding measurements.
[ ] Adjusting measurements to accommodate agile delivery.

Which of the following shows the register's value in sustaining benefits?

( ) All team members should use the register.
( ) Focus mainly in intangible benefits.
( ) Ownership and expected change achievement timing.
( ) Not capturing disbenefits.

12.2 RESILIENCY AND BENEFITS

This is an important topic for the future of work given the recent years' learnings and immense disruptions the world has experienced. Resiliency is the one quality differentiates you and allows you to succeed in the future. This also directly affects the practical driving towards sustaining benefits program team members should be involved in on a daily basis.

Key Learnings

- Rethinking what it takes to develop resiliency using the pandemic lens
- The importance of Benefits Harvesting
- Stakeholders create the necessary adaptability in the benefits agenda
- Using the HBR study to highlight why most companies have a gap between their strategic plans and need to create agility

12.2.1 Develop Resiliency Using a Pandemic Lens

Figure 12.2 is a strong reminder that the future program managers know how critical this capability is. They should practice adaptability in shifting the way of working and step up to turn tough threats into opportunities. Programs succeed when you have entrusted team members and contributors focused on value creation. Adjusting to changes fast requires an appreciation of the extermination value, reprioritizing quickly, while benefiting from digital. This gives the program teams the proactivity needed to deal with mega change events like a pandemic.

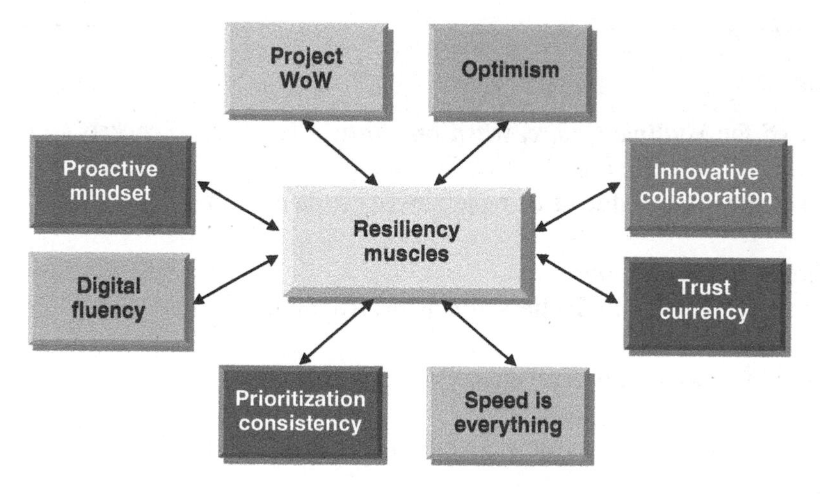

FIGURE 12.2 Resiliency Muscles

12.2.2 Benefits Harvesting

Dr. Harold Kerzner has emphasized across many of his publications, the criticality of the timing and ownership of harvesting. A key question to address regarding harvesting of benefits is to take care of the benefits transitioning and all the associated support necessary. Continual collaboration has to be planned for if future support for the long-term benefits achievement, in some cases, is vital too.

- The problem is not identifying the benefits or managing programs to create benefits
- The challenge is in managing the transition once the programs components are over
- The importance of continual collaboration

12.2.3 Stakeholders Adaptability

Skillset

- Change management
- Adaptive strategic realignment

Mindset

- Careful purpose analysis (due diligence, data)
- Outcomes focused (critical shift to achieving value to make sure the key stakeholders get this adaptability under pressure supportive in the achievement of change)

12.2.4 Closing Agility Gaps

T-shaped skills remind us that what we need the most of in tomorrow's programs, is more the cross-cutting change skills, closeness to customers, EQ, and clock speed.

- Staying close to customers
- Faster time to market
- Closer match between capabilities and opportunities
- Enterprise agile values (do we have the right culture to support and sustain agility)
- Lower risk (while having the right risk appetitive)

The So What:

- Benefits harvesting takes time
- Program managers need to adapt, have the right dialogues, and create benefits ownerships
- Mindset and skillset, T-shaped skills, managed degree of risk, and closing gaps in adaptability
- Develop resiliency to help you achieve and sustain benefits

Review Questions

Parentheses ( ) are used for Multiple Choice, when one answer is correct. Brackets [ ] are used for Multiple Answer, when many answers are correct.

_____ is an example of critical resiliency ingredients that program managers developed from COVID learnings.

( ) Leaders showing who is boss.
( ) Fluidity in the way of working.
( ) More F-2-F opportunities.
( ) Take more time to prioritize and decides.

(continued)

(continued)

Brené Brown said "There is no innovation and creativity without failure." Why is that critical mindset shifts for benefits achievement?

Choose all that apply

[ ] Providing more dependable structure.
[ ] The importance of collaboration.
[ ] The necessity of continual strategic realignment.
[ ] Work becomes more interesting.

Which of the following closes agility gaps?

( ) Focus mainly on programs agility.
( ) Becoming risk averse.
( ) Taking more time before releasing to market.
( ) Better alignment between capabilities and changes.

12.3 AN OWNERSHIP ENVIRONMENT MATTERS

Key Learnings

- Ownership is not policing
- Success requires openness about the right level of ownership
- Making change happen and achieving of benefits will fall apart without ownership

12.3.1 Criticality of Ownership

You bet it is critical!

- Proper value mindset is critical for the creation of the ownership environment
- The approach to ownership should cover: *the journey* from business case to sustaining benefits, collaborative transitions, time to verify achievement, and seamless transitions (why, what, and how have to be followed to ensure we have the right purpose)
- Ownership requires *responsible* stakeholders: Individuals who care about achieving value of a program's investment and who have managed to create commitment to joint focused outcomes (like in the case of Pat Lencioni and his five layers moving from trust to focus on the right results and getting the team to be willing to invest the time and effort to build the right ownership)

12.3.2 Secrets of Cascading Ownership

Like in the case of Booz Allen and their secret sauce discussed in previous chapters.

- Social leadership
- Collaboration
- Learning (need to be the value proposition creator)
- Empowerment
- Trust (this has to be achieved and we got to protect it to create the empowerment in the environment)

12.3.3 Timely BRM Design

- Starting with thorough identification (exactly what the benefits are)
- Establish ownership for harvesting of the benefits
- Build change management muscle (the heart of the matter in achieving success, OCM skills – organizational change management; like peer review, third party inputs, and other ways to reassess and learn)
- Clarity of assumptions (throughout the way)
- Collaboration setup (important step for ownership – mechanisms, environment, support, etc.)

12.3.4 Portfolio Linkages

Looking at the program and its benefits strategically and having a holistic view is fundamental.

Ownership is in the mind and the heart:

- Clear strategic focus (mindset)
- Motivation for value (heart)
- Portfolio adaptable structure (ways by which I can cascade)
- Strengthening horizontal working drivers (specific deliverables across program components while motivating the teams to work across the program boundaries to make sure this is achieved)

The So What:

- Ownership environment requires a secret sauce for cascading leadership, building trust, and true empowerment
- Consistency in ownership
- Raise the connection to the portfolio level
- Clarity of benefits and connecting the intellectual to the motivation to stay consistent with the design and plans for harvesting and owning benefits

Review Questions

Parentheses ( ) are used for Multiple Choice, when one answer is correct. Brackets [ ] are used for Multiple Answer, when many answers are correct.

_____ most supports the successful establishment of a benefits ownership environment.

( ) Stakeholders with a clear focus on profits.
( ) Having an approach that shows clear transitions.
( ) Focus on rewarding high performers.
( ) Detailed planning cantered on outputs.

(continued)

(continued)

What are ingredients examples mentioned in the lesson that directly support effective benefits ownership cascading?

Choose all that apply

[ ] Social Leadership.
[ ] Using one approach across the enterprise.
[ ] Directing.
[ ] Empowerment.

Why is it critical that the mindset and the heart critical to cascading ownership down and across?

( ) Illustrates clear hierarchy in the organization.
( ) Illustrates the ability to horizontally connect the programs.
( ) Makes stakeholders question value.
( ) Being consistent.

12.4 MANAGING FOR TRUST

Key Learnings

- Trust is key to overall success of the program
- It is foundational to how we communicate well
- Designing the right approach, engaging, and continually providing this building block for success

12.4.1 Control and Achieving Benefits

- This should not be confused with policing:
- Investing in the right metrics is worthwhile (like the Kerzner's maturity of metrics)
- Future of programs is value-centric (tell me how I am measured!)
- Protecting critical conversations matters (have the right attention on the bigger value topics with a model or a process that get the environment to be open to the tough dialogues – key role for the program manager and the benefit owner to create and support that)
- Success is in a **balance** of control and autonomy (making sure that I am entrusting and empowering the team to perform)

12.4.2 Speed of Trust[3]

Five Waves of Trust – Self, Relationships, Organizations, Markets, Society (different layers)
Behaviors – what matters the most for program teams

- Character Behaviors:
  - Straight Talk, Respect, Transparency, Humility, Loyalty (protects critical dialogues)

[3]Adapted from Covey (2008)

- Competence Behaviors:
  - Results, Improvement, Real, Expectations, Accountability (helps the solid foundation of ownership)
- Both:
  - Listening, Commitments, and Extending (building trust centered bridges and relationships)

Translate that to something that ties directly to your practice and how you work with your teams.

12.4.3 Trust Across the Program Team

- A trust-based program team ingredient:
  - Set up program team's values (program charter) – addressing the "why" questions
  - Delegating change ownership = clarity of outcomes + autonomy + support for open critical conversations (you got to protect those conversations)
- Role of program manager is to remove any barriers to connecting and to exercise strong people focus (this by itself is already a tremendous value to the program team – getting the team to proper focus)

> **TIP**
>
> The program manager is a trust builder and can use trust to create ownership and drive the critical program conversations

12.4.4 Trust and Focus on Benefits

Figure 12.3 summarizes key linkages between trust and the achievement of benefits. It also summarizes many of the capabilities of the program managers and the PMO that are valuable to achieving program outcomes. Trust is good – control is better, common theme around thinking especially of surprises that could affect the benefits, care about the steps needed to create the ownership for maintaining trust.

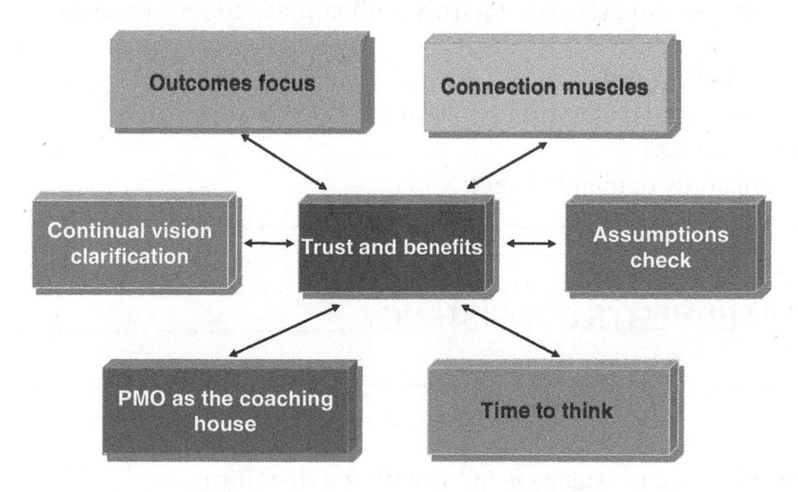

FIGURE 12.3 Trust and Benefits

The So What:

- Trust matters!
- Build the trust currency
- Find the right balance between control and trust
- Work on the mapping between the building of trust and the achievement of benefits
- Protection of critical conversations, the speed of trust, and the different behaviors and competencies of Program managers and the PMO maintain trust

Review Questions

Parentheses ( ) are used for Multiple Choice, when one answer is correct. Brackets [ ] are used for Multiple Answer, when many answers are correct.

_____ illustrates proper control is key to achieving benefits.

( ) Keeping executives informed of all program activities.
( ) Investing in building the right balance between control and autonomy.
( ) Focusing on scope, cost, and schedule metrics.
( ) Minimizing the amount of unnecessary dialogues.

Which of the following is a behavioral example that contributes to higher level of trust achievement? Choose all that apply

[ ] Exercising humility and admitting mistakes timely.
[ ] Apply fake it till you make it as necessary.
[ ] Avoiding straight talk and conflicts.
[ ] Extending trust across program teams.

Which of the following could best reflect a trust enabler that supports benefits focus?

( ) PMO strengthens level of control.
( ) Providing the program team an opportunity to think.
( ) Set assumptions once upfront.
( ) Key stakeholders constantly demand benefits.

12.5 CHANGE AND BENEFITS CONSISTENCY

Key Learnings

- Any time we talk excellence, consistency helps us with the right focus
- How can we consistently achieve the benefits expected out of a given change?

12.5.1 The Learning Organization

- Fish stick from the head!
- Driven by the Executive Team
- Focused on building the consistency muscle (exemplify that in my work as a program leader)
- Learning ingredients (e.g. NASA and the role of the Chief Knowledge Officer, working directly with the executives to create the right change):
  - Collaborative environment (must-have)
  - Willingness to experiment (e.g. different ways of delivery)
  - Value focused (have we involved and created ownership?)
  - Growth mindset (many organizations talk about this, yet we have to execute on this to provide the connective tissue for a strong learning organization – think of your program as the lab concept to extract the right value from your work)

12.5.2 Holistic Nature of Programs

See the forest from the trees!

Programs are natural in supporting consistency in achieving change results:

- Usually designed with a focus on effectiveness of resources usage
- A program lifecycle is centered on value creation (natural and key to the program management success given the reviews along the way)
- Program manager is expected to excel in integrating (imbedded in the program work)
- Multiple opportunities for continuous learning and improvement (exercise the discipline across the check points and with the right fitting governance to exemplify the learning to create the right consistency)

12.5.3 Benefits Across Initiatives Portfolio

Practices for Consistency across the portfolio of initiatives:

- Success = Disruption Handling (there will be surprises along the way)
- Seamless Changing Priorities Management (clarity of authorization and governance)
- Entrusting staff to drive program changes (empowerment ingredient)
- New ways of Diverse Inputs Inclusion (not a foreign topic when it comes to program work, need to create new ways to bring the mix of ideas in the discussion to enhance learning and consistency)

12.5.4 Benefits Realization Manager

Blue sky thinking of the roles helps in accomplishing change and achieving benefits.

Should we have a person with that title in our programs and across the business?
The Voice of Benefits Realization Ownership – high focus on the realization
Stewardship of Change culture – the shift needed

- supported by growth-minded leadership
- encourages and coaches the stakeholders around change benefits (mostly a missing component that is an easy fix)
- accountable for creative inclusion (massive inclusion of views, even opposing ones, to strengthen our learning – a dynamic and ongoing process)

This is such a simple notion that is foundational to organizations that want to excel in learning.

The So What:

- Growth mindset, commitment of leadership, and cascade trust building
- Holistic nature of programs creates wonderful learning opportunities and stretching
- Unique roles for creating the muscles for learning and the right ingredients for looking at things at the portfolio level
- It is a big yet doable responsibility, being a benefits manager, that creates a direct difference in your success

Review Questions

Parentheses ( ) are used for Multiple Choice, when one answer is correct. Brackets [ ] are used for Multiple Answer, when many answers are correct.

Which of the following is a healthy ingredient in the design of a learning organization culture?

( ) Rolling out one size for the organization.
( ) Hands-off executives.
( ) Trying something new every time.
( ) Growth mindset.

What is unique about programs in their holistic nature's support for benefits realization? Choose all that apply

[ ] Gates provide strong governance opportunities.
[ ] Focused on effective resourcing by design.
[ ] Program manager is a hero.
[ ] Program manager is good at integrating.

Why is the Benefits Realization Manager a possible strong asset for future programs?

( ) Rewarding command and control leadership.
( ) Making all key decisions.
( ) Coaching stakeholders around benefits.
( ) Ensuring most vocal opinions are considered.

REFERENCES

Letavec, C. (2004). *Strategic Benefits Realization*. J. Ross Publishers.

Melton, T., Iles-Smith, P., and Yates, J. (2008). *Project Benefits Management; Linking Projects to the Business*. Butterworth-Heinmann Publishers, an imprint of Elsevier Publishers.

Covey, S.M.R. (2008). *The Speed of Trust*. Simon & Schuster.

CHAPTER 13

Change Scientists

This chapter is focused on the role of change scientists and how data and trends enhance decisions and enable future change success. Change management is such a complex topic and in many cases, it spills into organizational changes, and even other implications on business models. For programs to deliver the anticipated benefits and outcomes, we need to have fluid access to data to enhance governance, accelerate speed of decisions, and give us the ability to tilt fast as the conditions change. It is the intent here to potentially see the change scientists helping in filling this important void.

The ***Program Way***, led by many of these change scientists, could be a result of changes in the business model, or it requires changes to the business model. As programs develop new products, the result can be changes that are necessary in the organization's business model. Change scientists use their collaboration skills, data analytics, and the power of Artificial Intelligence to simulate what the future organization needs to shift to, and what changes in how success is viewed, results in specific actions that they should drive.

Key Learnings

- Understand how the Change Scientists revolution is linked to a closer focus on this power skills.
- Learn the attributes of the change scientist that are critical ingredients for change success.
- Explore the principles of creating the right design and mix of program metrics.
- Understand the important tools for decision making and the use of data and trends to expedite and enhance decisions.
- Develop a strong awareness of the change scientist's value for enterprise decisions and its impact on the program's core team way of working.

Program Management: Going Beyond Project Management to Enable Value-Driven Change, First Edition. Al Zeitoun.

13.1 THE CHANGE SCIENTISTS' REVOLUTION

Key Learnings

- The art and the science mix to make change happen
- The need for a combo of skills between leadership, data, engaging, and collaboration
- The change scientist's role

13.1.1 The Man in the Arena

The following supports the context behind the criticality of the role of change scientist:

- *Theodore Roosevelt*: "The credit belongs to the person in the arena – if he fails, he fails daring greatly" (experimentation, failing, learning and many of the other attributes that are linked to our success)
- *Brené Brown*: correlation to creating as change scientist
  - Vulnerability is about showing up and being seen
  - If you are not in the arena, I's not interested in your feedback (changing how we work in the organization centers about questioning, continuously improving and learning – focusing on the ones with the right understanding of what is in the trenches)

13.1.2 Change Skills

Source: 489327 / 3 images.

As highlighted in this picture, with the energy and the passion in a carnival, the role of the change scientist is centered on driving the energy for transforming and positive change.

- Way of Working (WoW) anchored in change (affecting consistency and the right dialogue, risk-based governance)
- Prototyping
  - Fail fast and inexpensively
  - Upfront in design thinking (quickly bring ideas to the mix)
- Design Thinking Strategy
  - Focus on changing needs (dealing with the multiple changes)
  - Idea generation creativity (quick feedback and learning)
  - Complex connections clarity (empathy)
  - Quick small bets (quick wins, minimal viable products (MVPs), allows you a better feel of the place and the resulting change)

The So What:

- Setting the foundation for change scientists
- The man in the arena (entrusting key people for feedback)
- Building the skills of this critical change scientist role

Review Questions

Parentheses ( ) are used for Multiple Choice, when one answer is correct. Brackets [ ] are used for Multiple Answer, when many answers are correct.

_____ illustrates the person creating change is ready to be truly vulnerable.

( ) Consistency, repeatability, and control measures.
( ) Willingness to roll up the sleeves and be in the trenches.
( ) Focus on providing improvement feedback.
( ) Value, strong alignment, and communications.

Which of the following are key ingredients of a Program Manager driving change? Choose all that apply

[ ] Toughness with people.
[ ] Ability to take the hits on behalf of the team.
[ ] Ideating as much as possible.
[ ] Falling daring greatly.

Which of the following is an example of design thinking advantages?

( ) Rushing into prototype.
( ) Overthinking the information at hand.
( ) Turing complex connections into clear maps.
( ) Expecting final solution fast.

13.2 THE POWER SKILL FOR PROGRAM SUCCESS

Programs are about how we excel in the balance required for achieving change.

Key Learnings

The best scenario – the program managers are capable change scientists
Putting together the complementary program team that covers this power skill demand
Reviewing the key mindset shift in realizing benefits between a waterfall and an agile approach
Building change pulse and collaboration routines

13.2.1 Program Managers as Change Scientists

- Business Context (matters for program success)
- Adaptable Approach (need to close organizational agility gaps)

- Data Analytics (Trends, Power Business Intelligence (BI))
- Intentional Communications
- Strategic Alignment (how can we make sure that we are strategically linking change to the objectives of the business)

13.2.2 Complementary Team Skills

- Conducting a detailed inventory of the teams' power skills (previously referred to as soft skills, yet they are the true key ones affecting the achievement of the change)
- Using a team-driven approach to reach consensus on needed skills mix
- Test the skills mix to achieve a change program mission (as a PM you got to audit and find out the kinds of shifts that are needed in the skillset)

Figure 13.1 shows the maturing of the PMI Talent Triangle skills groups. This is likely to continue to change and expand in the future. A change scientist must exit a good balance of these groups of skills.

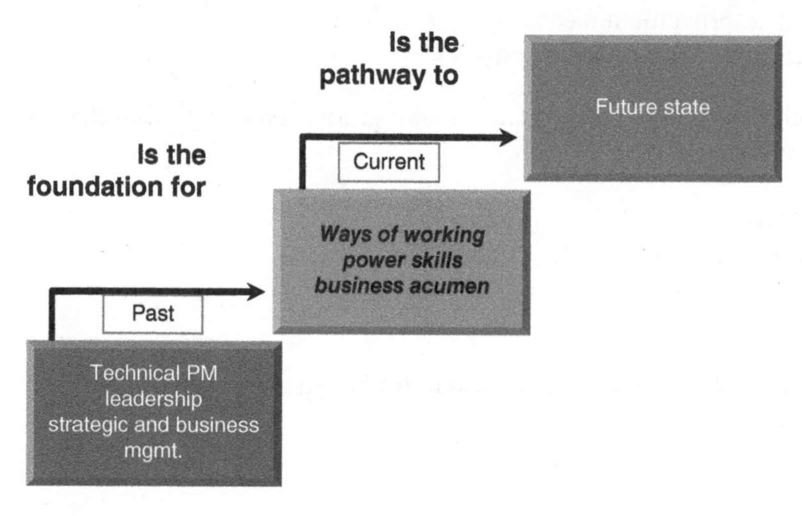

FIGURE 13.1 Power Skills Maturity
Source: Adapted from PMI Talent Triangle

13.2.3 The Mindset Shift

Figure 13.2 shows an example of the mindset shifts that change leaders create. Many of the choices shown are affected by the highlighted factors, such as trust level. It is important for the leaders of change and the change scientists to study and understand the impact these factors could have on supporting the aspired changes.

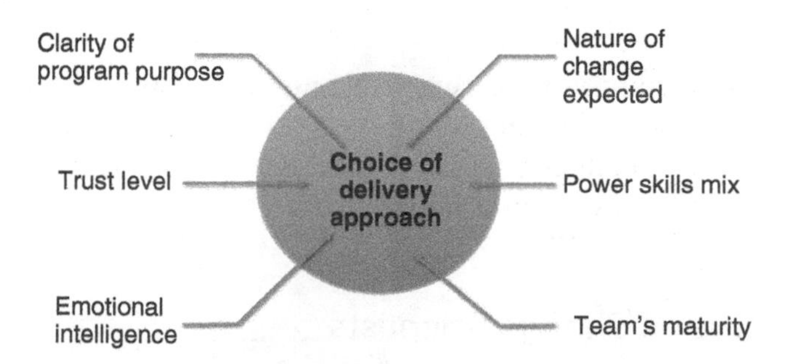

FIGURE 13.2 Adaptability Around Delivery Approach

Mindset shift towards: Knowing what are we after, keeping our eyes on the ultimate change outcomes, asking the questions, addressing tough topics, dealing with politics with resilience, maintaining the proper trust level and opportunities for the team to have a voice, and collaborating as necessary.

13.2.4 Change Pulse

- Customized hybrid interactions (finding ways to pulse across settings)
- Continual sensing of change direction (the right health checks)
- Adjusting the communication plan to adapt to change gaps
- Change ownership validation (like in the case of the benefits realization manger and others – ensuring no gaps in taking the change across the finish line)

The So What:

- Making sure that the Power Skills dominate across the team
- High focus on business acumen and mindset shift
- Applying a change pulse to adjust course and using data to get things back on track

Review Questions

Parentheses ( ) are used for Multiple Choice, when one answer is correct. Brackets [ ] are used for Multiple Answer, when many answers are correct.

_____ illustrates the person creating change is ready to be truly vulnerable.

( ) Consistency, repeatability, and control measures.
( ) Willingness to roll up the sleeves and be in the trenches.
( ) Focus on providing improvement feedback.
( ) Value, strong alignment, and communications.

Which of the following shows the team attributes relating to the mindset shift necessary for a successful hybrid delivery approach?

Choose all that apply

[ ] Power skills mix.
[ ] Intellectual Quotient.
[ ] Trust level.
[ ] Program's budget.

Which of the following is an example of the need of a hybrid interaction process?

( ) Most of the team is co-located.
( ) Program team has shown consistent use of waterfall.
( ) Management wants to send a cultural message.
( ) A mix of program team sites and generations.

13.3 THE PROGRAM METRICS MIX

Key Learnings

- The need for change scientists
- Value metrics
- Overcoming barriers

13.3.1 The Need for Change Scientists

The governmental agencies across the world, typically with a diverse and strong mission, are great examples for the need of this role. Many of these agencies have common ingredients as shown next that would make the need for an effective change strategy, supported by the right leading roles, a fertile ground for successful implementation cases.

- Complex mission crossing diverse stakeholders
- Enterprise portfolio in Billions of Dollars (affecting people's lives and thus transformations are critical)
- Transformation of systems and services
- Change readiness variances (big priority and focus whether in a nuclear program or a simple manufacturing operation on a shop floor to address any variances in our change readiness)
- Data-supported change

13.3.2 Holistic Value-Centered Metrics Mix

Dr. Kerzner's points, as highlighted by figure 13.3, around the move to strategic value metrics, set up the importance of this topic and how a holistic view of metrics would help the program team get the right focus on what matters for achieving change outcomes.

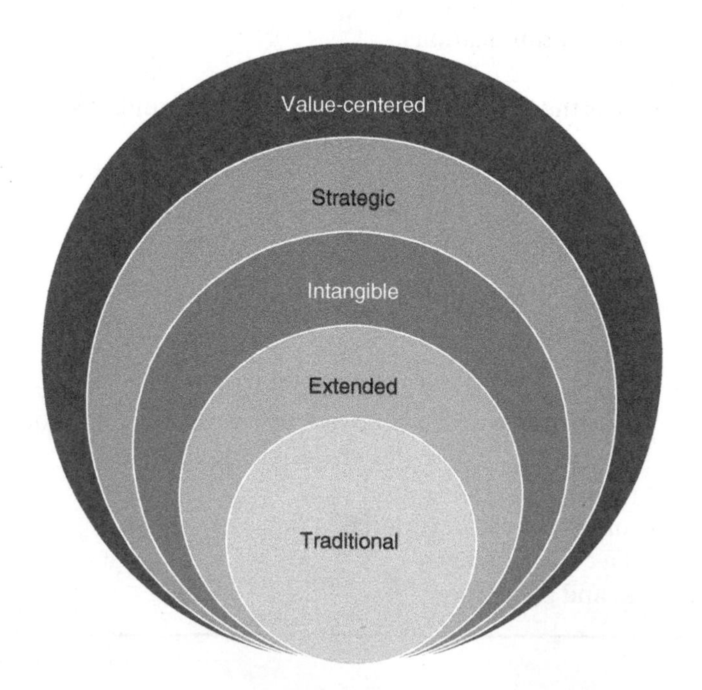

FIGURE 13.3 Evolution of Metrics

Metrics Shift

1. The shift ties closely to the program discipline and maturity (as was highlighted in the maturity path by Kerzner)
2. Program manager needs the backing of executives
3. Value-centered ensures metrics that matter to program's success (ownership required and mindset shift needed)

13.3.3 Barriers and Biases

- **Bias**
  - Extreme action or perfection focus (both could hinder the chances of focusing on what matters)
  - Preference for executive views
- **Barriers**
  - Tell me how I am measured (results in where performance is focused on)
  - Short-term view (need to think of the path – long-term view)

The So What:

- Strategic metrics are needed
- Care should be exercised around taking biases and barriers out of the way
- Program managers should ensure that we have the right focus on what matters
- Critical conversations should take place and change leaders should working with benefits owners and executives

Review Questions

Parentheses ( ) are used for Multiple Choice, when one answer is correct. Brackets [ ] are used for Multiple Answer, when many answers are correct.

_____ is an example of what brings the science and people side of change together for effective outcomes.

( ) Focus on profit.
( ) Data analytics use.
( ) Deep technical expertise.
( ) Giving program teams full autonomy.

What are good examples of the shifts towards value in designing our change metrics? Choose all that apply.

[ ] Ability to include intangibles in the metrics mix.
[ ] Developing the maturity of the program team.
[ ] Become popular with management.
[ ] Rethink the strategic implications of change.

(continued)

(continued)

What organizational barrier could limit the full value of program metrics?

( ) Perfection.
( ) Focus on the most senior experts views.
( ) Comfort with what the boss thinks.
( ) Not considering the long-term effect.

13.4 DECISION-MAKING MASTERY

Is there such a thing as true mastery? Decision-making excellence is targeted here. Gets into creating a critical muscle, quality of decisions, data, and multiple stakeholders' voices.

Key Learnings

- Manage a Veteran Affairs portfolio with a benefits-centered approach for prioritization decisions at the highest level of that agency
- The UAE Minister of Happiness as a top leadership example in focus on a wider programs' benefits value impact
- Models of decision-making and the advantages of clear graphical decisions displays such as Decision Trees

13.4.1 Prioritization

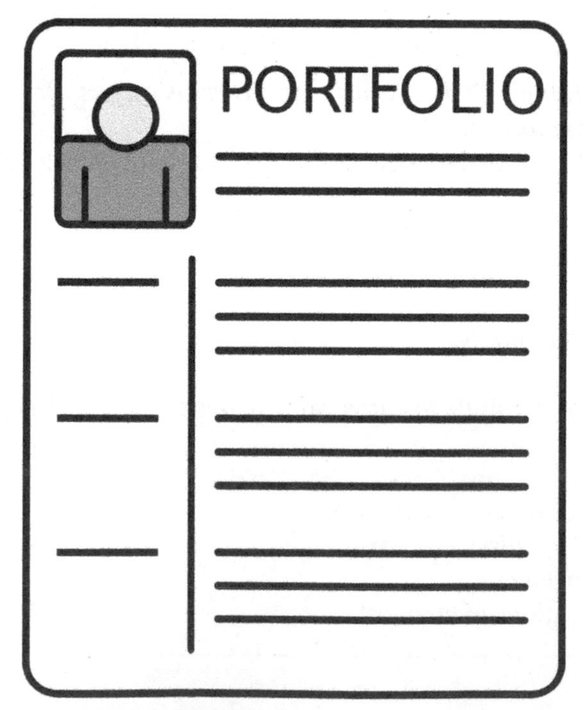

Source: TKaucic / 17 images.

- Decision mastery is about strengthening the decision-making muscle (requires investment)
- Governmental organizations as political environments, yet most impactful missions (can't be taken lightly and thus a higher need for effective prioritization)
- Lot of data across initiatives with a need for continual prioritization
- Move to strategic dialogue and value increase chance of benefits-centered resourcing
- Role of leadership in culture creation
  - asking right questions and using relevant dashboard strategic metrics
  - risk-driven prioritization (at the enterprise level)
  - rewarding open dialogue (enabling better decision-making)

13.4.2 Leadership Example

Source: Clker-Free-Vector-Images / 29538 images.

- Happiness as a country strategic goal
- UAE Ministry of Happiness value proposition is promoting optimism and positivity (many of the attributes change scientists promote)
- Happiness meter, champions, training, and a cascading structure across organizations (ideas to supporting excitement around critical decisions)

13.4.3 Models of Decision-Making

- Data-rich and value-driven (Expected Monetary Value – EMV, putting a weight on the decision)
- Are the insights from background stories helpful in creating decisions clarity? (Balance gut and data in the mix)
- Decision trees allows clarity of decisions path and enables capturing the expected value associated with decisions tradeoffs (e.g. take this job or that, follow this solution or that, see the path, right dialogues and players, consistently brining us back to true program value)

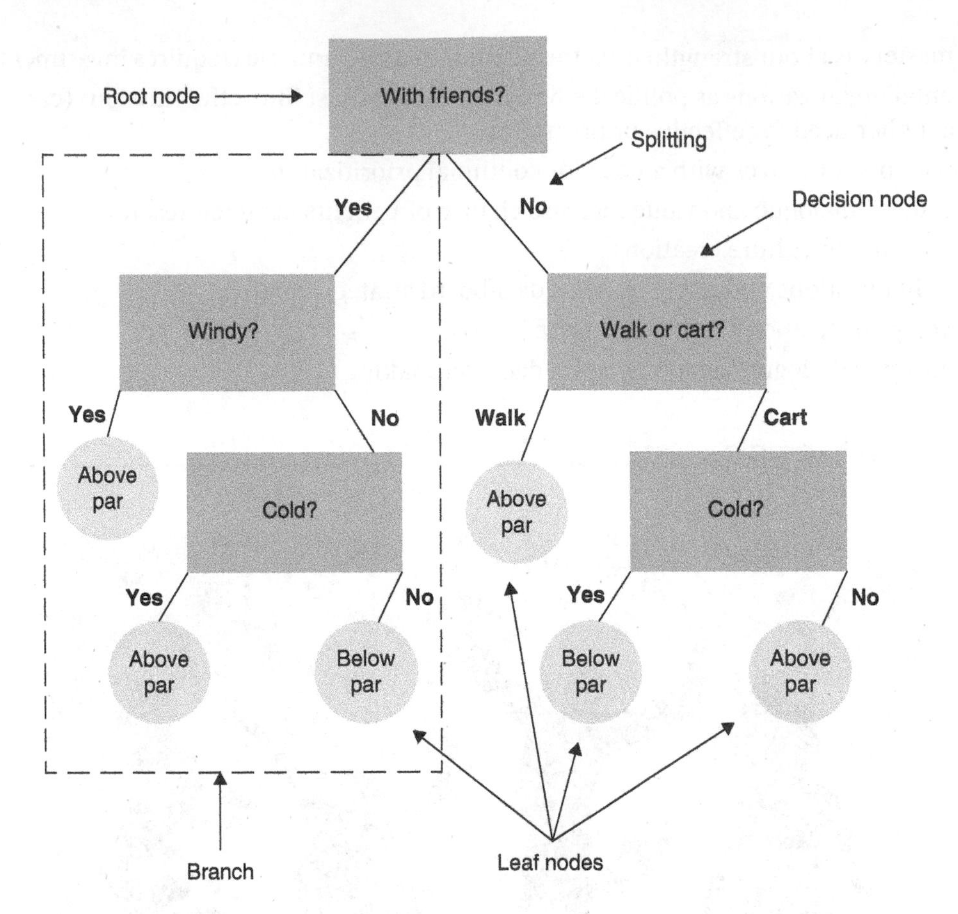

FIGURE 13.4 Decision-Making Tree Example
Source: https://www.mastersindatascience.org/wp-content/uploads/sites/54/2022/05/tree-graphic.jpg

Figure 13.4 shows an example of how decision trees, among other tools and visual ways to articulate and interpret decisions, could help with the alignment of the program stakeholders on the direction to take and the reasoning behind that. These could be all small steps towards some level of mastery.

The So What:

- It is a journey, a continual exercise
- Happiness meter
- Tools that help decision-making, e.g. decision-tress and EMV
- Core is strengthening our prioritization to increase the chances of change achievement

Review Questions

Parentheses ( ) are used for Multiple Choice, when one answer is correct. Brackets [ ] are used for Multiple Answer, when many answers are correct.

_____ contributes to enhancing our programs prioritization around value.

( ) Opinion of most senior stakeholder.
( ) Risk-driven prioritization.
( ) Spending more time in analysis.
( ) Gut feel.

Which of the following enhanced the UAE's success in spreading the value of happiness in the country? Choose all that apply

[ ] Talking about happiness in every national event.
[ ] Having a CEO of happiness in various government entities.
[ ] Culture enhancements with the right supporting models.
[ ] Penalizing unhappy citizens.

Which of the following is an example of a tool that simplifies the representation of decisions alternatives?

( ) Resource diagram.
( ) Work breakdown structure.
( ) Decision trees.
( ) Gantt charts.

13.5 THE PROGRAM CORE TEAM CHANGES

Key Learnings

Change scientists need the core team and that they believe in the changes in approach and focus they drive. Bottom line is utilizing the core team to instill the changes and proceeding with the tough calls required.

13.5.1 Organization's Benefits Strategist

- Focus – line-of-sights
- Forecasting – risk management
- Systems – holistic
- Transformational – EQ (emotional intelligence allows us to relate well, manage stress, build alliances, and thus becoming true strategists)

As shown in Figure 13.5 and highlighted earlier, these are attributes that affect the role of the program manager and change scientists.

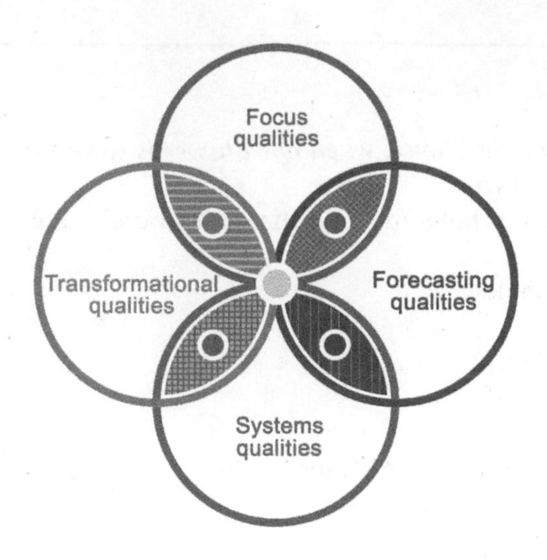

FIGURE 13.5 Benefits Strategist Attributes

13.5.2 Change Scientists Workstream

Picture should remind us of the criticality of creating a dedicated change workstream.

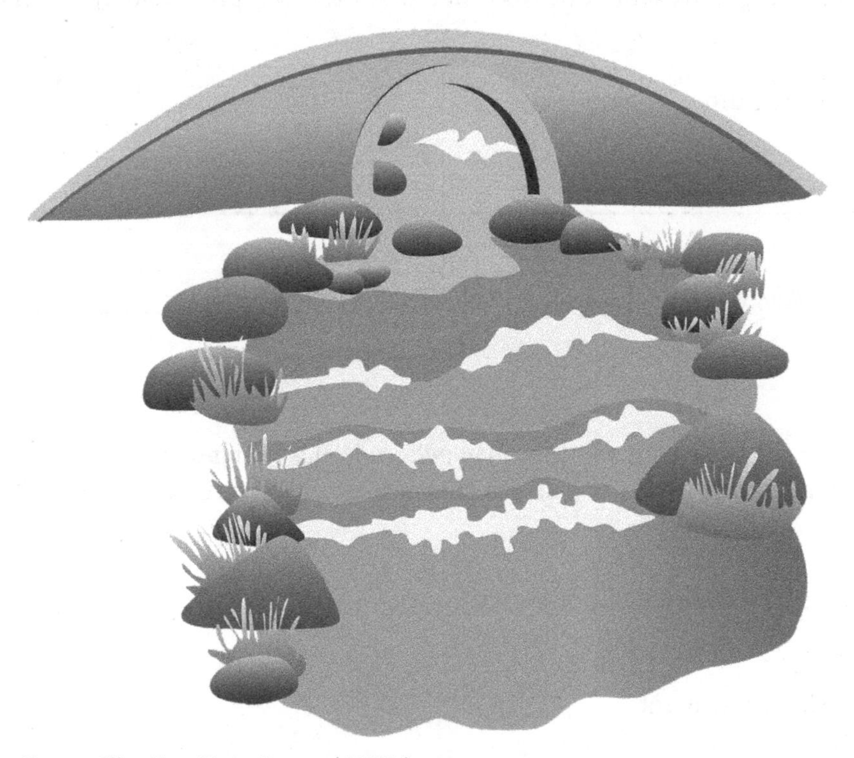

Source: Clker-Free-Vector-Images / 29538 images.

- **Focus:**
  - Cut through distractions and maintain focus on change (dedicated to making change happen)
- **Change Readiness:**
  - Roll up your sleeves and work with the target team (assessments, pulse checks, using the man in the arena concept)
- Change Stickiness:
  - Continual engagement to sustain changes (past getting there, this core team ensures change sticks)

13.5.3 Change Stories

13.5.3.1 Ingredients

As shown in Figure 13.6, inspiring change, how we tell the story, like a running river that flows with the proper cadence, points of views, perspectives, lasting impact, and ultimately getting us to the stickiness required.

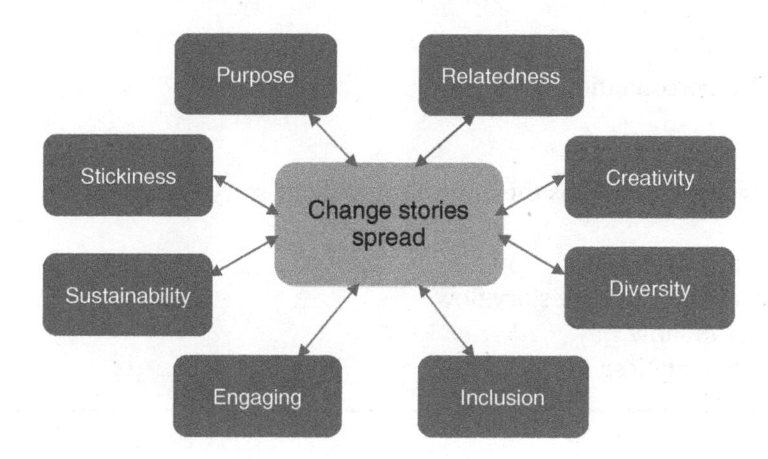

FIGURE 13.6 Powerful Change Stories

13.5.4 The Coach for the Core Team

- The changing role of the program manager (using the PM potentially as a coach)
- The responsibility for core team's attributes development (PM connects the dots to ensure the change happens)
- Ongoing collaboration between program managers and their core teams to sustain changes (openness, integration role of the PM, continual journey focus)

The So What:

- Dedicated workstreams help us achieve stickiness
- Role of program managers continues to change
- Focus-readiness-stickiness balance
- Core team keeping its eye on the ball
- The value of the entire benefits map

Review Questions

Parentheses ( ) are used for Multiple Choice, when one answer is correct. Brackets [ ] are used for Multiple Answer, when many answers are correct.

_____ is the attribute that enables the program manager to get the change team focused.

( ) Collaboration.
( ) Line-of-sight.
( ) Transformation qualities.
( ) Dynamic management support.

(continued)

(continued)

Which of the following is a key advantage of having a dedicated change workstream in securing change outcomes?

Choose all that apply

[ ] Fun environment.
[ ] Readiness check.
[ ] Dealing with troubled personalities.
[ ] Stickiness.

Which of the following strengthens the impact of a change story?

( ) Making it general across audiences.
( ) Including the target audience in the story flow.
( ) Sharing information the same way.
( ) Depending mainly on analytics.

CHAPTER 14

Adaptable Roadmaps

This chapter focuses on the adaptable change roadmaps and how diversity of views and cocreation strengthen benefits traceability and stakeholders' satisfaction. It is important to understand the impact of when program roadmaps are designed and adapted around value. Adaptability is a key quality in future program roadmaps to address the high degree of complexity and uncertainty encountered in program work. Rediscovering how we plan, adjust, and refine our roadmaps requires a learning appetite on the part of program leaders and team members. This pays off in enabling the creation of relevant and guiding roadmaps for programs' success.

Key Learnings

- Get immersed into the importance of closely sensing stakeholders' needs and the balancing of the fitting light governance
- Learn the critical importance of cocreation to secure shared ownership
- Learn how to value and consistently practice extracting diverse perspectives to support traceability and securing ownership of benefits
- Develop a strong appreciation of how cultural understanding enables critical ownership and contributes to the success of adaptable leadership practices

14.1 VALUE-BASED PROGRAM ROADMAPS

Key Learnings

- Building on the ideas of the change scientists and the core team
- The foundational enablers like the roadmaps that help us adapt and stay focused on achieving change
- Making sure that in what we design as roadmaps, the value-based ingredient is increasing the effectiveness of the use of these roadmaps

Program Management: Going Beyond Project Management to Enable Value-Driven Change, First Edition. Al Zeitoun.

14.1.1 Program Roadmaps and Value

Source: Clker-Free-Vector-Images / 29538 images.

Success could be our worst enemy – could lead to resisting change.
If roadmaps are not designed with a focus on value:

- Programs will lose their effectiveness in achieving stakeholders' satisfaction
- Gaps will widen between strategy development and implementation (huge point affecting success)
- Design will not get the buy-in if value is not clear and collaboration was not established (it's about ideas coming together and working across stakeholders)

14.1.2 Stakeholders' Traceability

- Brené Brown's view of ownership versus blame (lovely reminder to control our pulses to rush to blame)
- Multiple stakeholders, confirming the need for clear roles and responsibilities (need to fill any of the gaps)
- Benefit owners and/or program sponsors with the keys to measuring achievement
- Design for both tangible and intangible value (as in the Kerzner metrics maturity recommendations)
- Continual mapping to strategy (Have the right strategic dialogues taken place?)

14.1.3 Value-based Organization

- Rethinking what value means
- Ensure meaningful success achievement (e.g. in the consulting space and checking the box on deliverables is not meaningful enough)
- Faster failing and learning
- Creating spaces for co-creating (huge across ways of working)
- Decision-making and prioritization muscles (reminder of strengthening our prioritization muscles)

The So What:

- Looking at the design to be value-centered
- Trace the stakeholders' ownership
- Continue to build on the learning-based organizations to become value-focused
- Make sure we have the right supportive organization in its design and its support to the critical programs

Review Questions

Parentheses ( ) are used for Multiple Choice, when one answer is correct. Brackets [ ] are used for Multiple Answer, when many answers are correct.

What could happen if roadmaps were not designed with a value orientation?

( ) Closing the gap between strategy development and implementation will be a challenge
( ) Program success will be more focused
( ) Program managers will be blamed
( ) Stronger buy-in

Which of the following stakeholders has the ultimate responsibility for measuring the achievement of benefits?

( ) Program manager
( ) All stakeholders
( ) Benefits owner
( ) Program team

How can an organization sustain its focus on value?

( ) Stick to one view of program value
( ) Centralize interaction with customers
( ) Add more review milestones
( ) Question what value and success mean for different stakeholders

14.2 THE ADAPTING FACTOR

Adapting continues to be one of the most talked about topics in the era of continual and fast change. An adapting meter could be a way by which we measure our maturity as organizations and program teams in how well we adapt. This builds on the resilience practices that are critical for program managers. Having the right organization setup that supports adaptability is a key responsibility of the executive team and the organizational design architects.

Key Learnings

- Why optimizing strategy execution matters?
- Using strategy execution focus as the ultimate measure for BRM culture success
- Building adaptability
- Designing BRM as the connective tissue to value reporting

14.2.1 Strategy Execution Matters

Source: JESHOOTS-com / 142 images.

- Embracing focus (like in the case of the chess game)
- Providing the guardrails for the change roadmaps
- Aligns the program teams yet leaves the room for creativity and inclusion (adaptable roadmaps require fine balance)
- Ultimately, it is where value is seen and delivered (how we execute is a reminder why roadmaps are useful)

14.2.2 BRM Culture Success

Figure 14.1 shows an illustration that connects to the change scientists' attributes and the earlier ideas we discussed about the healthy culture needed to support effective change.

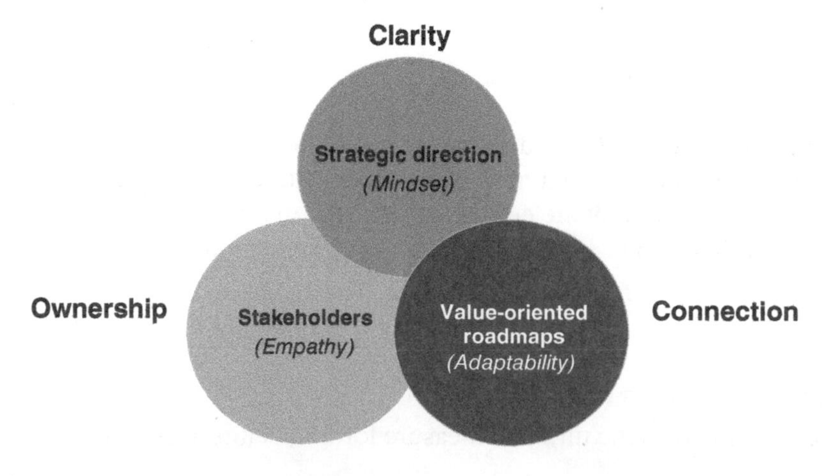

FIGURE 14.1 BRM Culture Attributes

14.2.3 Building Adaptability

The higher the uncertainty, the higher the expected value of adaptability on programs' success:

- Adaptability is everything in being ready to learn and change (learning organization)
- Rewarding the ability to bounce back quickly (cherish those qualities)
- The PMO as the driver for sharing adaptability success stories (powerful stories as a core skill)
- Role of benefits owners in connecting the dots (build that intentionality of roles)
- Coaching the value of agility, creativity, and strategic focus

14.2.4 BRM as the Connective Tissue

Source: geralt / 25607 images.

- Reporting = Opportunity for ***meaningful dialogues***
- **Mindset Shift**
  - Continual Value Check Ins (repeatedly as in some of the agile practices)
  - Business owners' connecting responsibility
  - Learning-driven behaviors (key focus)
  - *Simple rules*: ***relentless re-prioritization***, open dialogues, and risk-based decision-making

> **TIP**
>
> Program managers should make sure that we have the right adaptable roadmaps for the team

The So What:

- Right culture, right learning organization, right roadmaps
- Entrusting that the stakeholders
- Understand the importance of the adaptive factor to be ready for the kind of changes ahead of us

Review Questions

Parentheses ( ) are used for Multiple Choice, when one answer is correct. Brackets [ ] are used for Multiple Answer, when many answers are correct.

_____ is an example of how an adaptable roadmap contributes to creating a change culture success.

( ) Always ensure one strategy stays the same across the FY
( ) Building the roadmaps with and adaptable focus on value
( ) Spread ownership across all stakeholders
( ) Escalate as much as possible to management

Which of the following contributes to building adaptability in the program team? Choose all that apply

[ ] Showing a commitment to one methodology
[ ] Rewarding mechanisms
[ ] Using the PMO to build the relevant adaptability examples
[ ] Strengthening reporting focus

Which is of the following indicates that BRM alters the behaviors surrounding value reporting?

( ) Random value check-ins
( ) Quickly pointing to who to blame for errors
( ) Reporting becomes an opportunity for meaningful dialogue
( ) Gut-based decision-making

14.3 BALANCING GOVERNANCE WITH TRACEABILITY

Program teams should be able to trace where the value was extracted, where the value achievement happens short and long term, and then trace that in a way that could help the team learn and not police – true value of adaptive roadmaps. This also supports integrating performance and value.

Key Learnings

- Value motivates performance
- Integrated formula covers strategic performance indicators cascaded down to specific program changes (like in the case of a strategy house and how the elements link)
- Changing focus of performance reviews supports the continual learning and motivating value achievement (fundamental part to balancing performance)
- A coaching environment establishment increases the potential of an effective program manager role (PMs are ready to support the program team in continually shifting toward value)

14.3.1 Metrics Growth Ladder

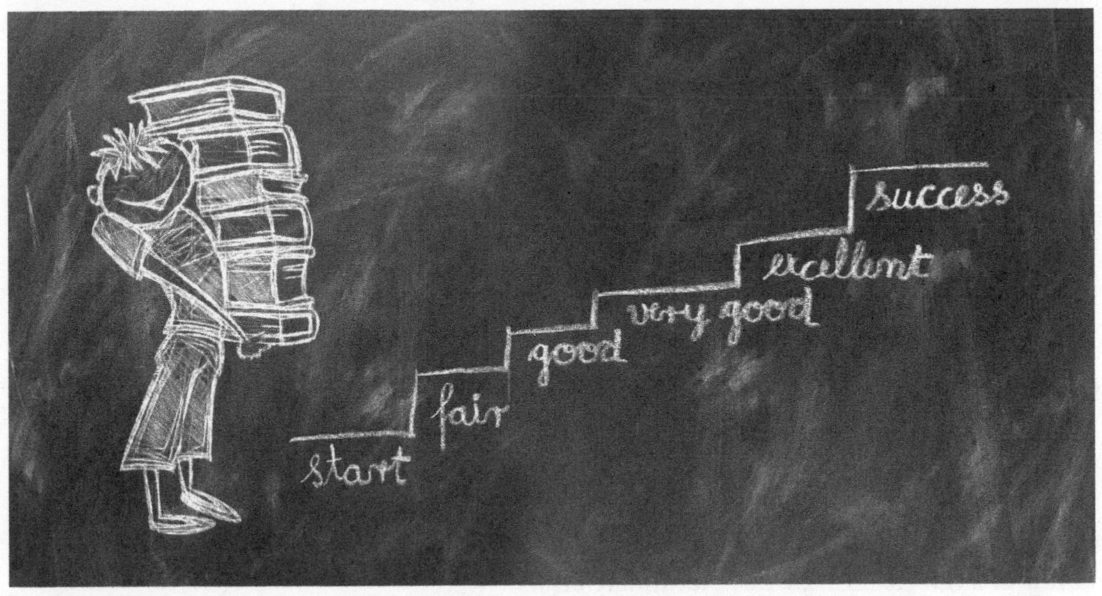

Source: geralt / 25607 images.

Stretch beyond value to intangibles and ultimately strategic value metrics

- Kerzner highlights the importance of moving toward strategic metrics
- Selecting metrics that matter starts about mid-way in maturity with a value focus
- Need to educate program stakeholders on the importance of intangibles (can't make the move or shift to strategic until intangibles are part of the mix and this is a strong sign of maturity, e.g. happiness score, morale.)

14.3.2 Governance and Traceability

Proper governance breaks down barriers to traceability, as highlighted by Figure 14.2:

- Entrusting the PMO and core program teams
- Creative collaboration and dialogues
- Energizing with capabilities, tools, and processes
- Timely sharing of uncomfortable news (critical element)
- Data-driven

TIP

Governance success is traceability centered, allowing us to trace the value and improvements we are creating as a result of programs' work.

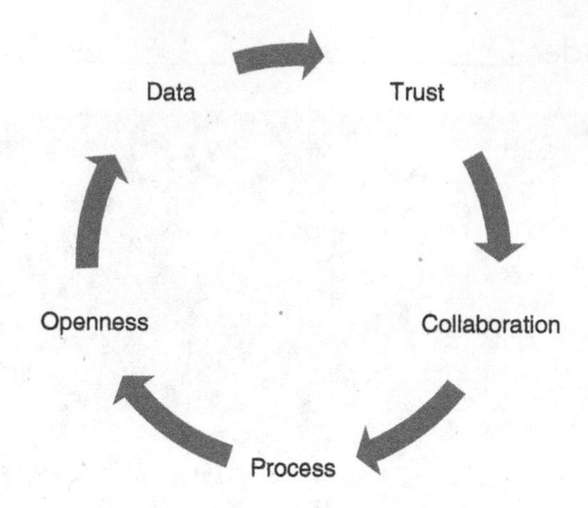

FIGURE 14.2 Traceability

The So What:

- Ongoing journey to improvement and finding the right fitting governance level
- Right mix of strategic metrics, need for intangibles
- Trust building, use of data, continuously improving and learning
- Ensuring roadmap is adaptable and balanced enough, bringing performance and behaviors in the mix to encourage and motivate the team

Review Questions

Parentheses ( ) are used for Multiple Choice, when one answer is correct. Brackets [ ] are used for Multiple Answer, when many answers are correct.

______ is an example of how performance could be integrated with value.

( ) Directing the team to focus on outputs
( ) Strategic indicators cascaded down to program changes
( ) Program managers have their home in the PMO
( ) Using enforcing style in performance reviews

Which of the following are signs that the organization has matured in the measuring of its program changes?
Choose all that apply

[ ] Increased investment in operational metrics
[ ] Use of intangibles
[ ] Drive the metrics by the PMO's agenda
[ ] Adapting to link to strategic outcomes

Which is of the following is a contributor to improving traceability?

( ) Sticking to sharing good performance news
( ) Being boss-centric
( ) Data-driven governance
( ) Use of a one-size-fit-all methodology.

14.4 COCREATED ROADMAPS

Could be talked about anywhere and often. It is about the stickiness of the change as we discussed. It is where the program team gets change commitment and buy-in. It is the magic sauce!

Key Learnings

- Using servant leadership to build the voice of the program teams
- How the Erin Meyer work in her culture map across world regions can be used to understand cocreation of roadmaps?
- Building the roadmap creation environment, gaining feedback, and opening critical conversations
- Use strong benefits sponsors to help equip future program managers in comprehending the capture of diverse perspectives

14.4.1 Servant Leadership

Figure 14.3 highlights critical qualities that the servant leaders could contribute to becoming more adaptable under the buckets of inspiring, coaching, and supporting. Like what Simon Sinek highlighted in "Leaders speak last," many important principles to set the cocreation foundation up. The program leader has a big role to play in here.

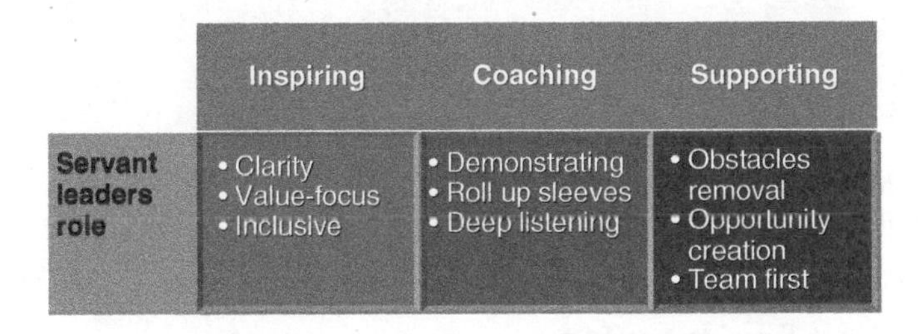

FIGURE 14.3 Servant Leadership

14.4.2 Culture Maps

As Erin Meyer did in her work across cultures, an example of which is shown by Figure 14.4, and looking at the opposite ends of spectrums as in dealing with confrontation and other attributes, this gives us another example for the enablers supporting adaptability:

- A safe place to share team members' voice
- Capitalizing on the strengths of various cultures (across globe, regions, etc.)
- Empathy drives creation of better program roadmaps (relating to different cultures generate much better cocreated roadmaps that are committed to)

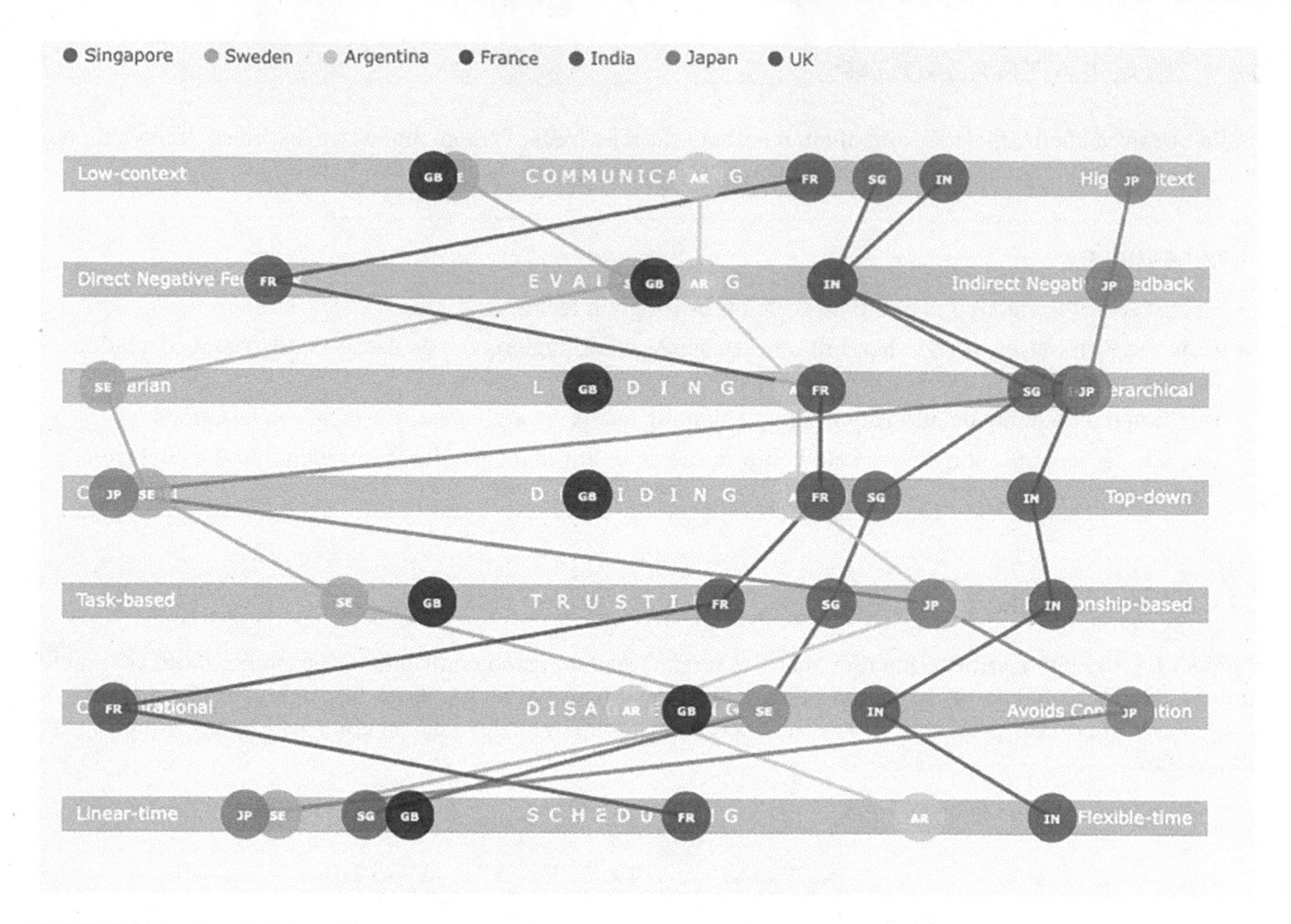

FIGURE 14.4 Sample Culture Map
Source: Adopted from the work of Erin Meyer as a sample Culture Map

14.4.3 Roadmap Creation Environment

Source: geralt / 25607 images.

- Operating as a coach (fun environment)
- Encouraging and rewarding critical conversations
- Higher risk appetite
- Collaboration focus (100% in the focus zone)
- Encouraging the use of feedback to grow roadmap commitment (nonstop focus)

14.4.4 Benefits Sponsors

As a PM use your executive leadership to help you ensure that the right supportive culture is in pace. Work with the key owners of value to ensure that the adaptability needed for achieving success is in place. Benefits sponsors are extremely valuable parties in the mix of what would ensure the program roadmaps are meaningful. These sponsors could help in bringing the diverse perspectives, highlighted by Figure 14.5, to the surface and thus strengthening the program outcomes.

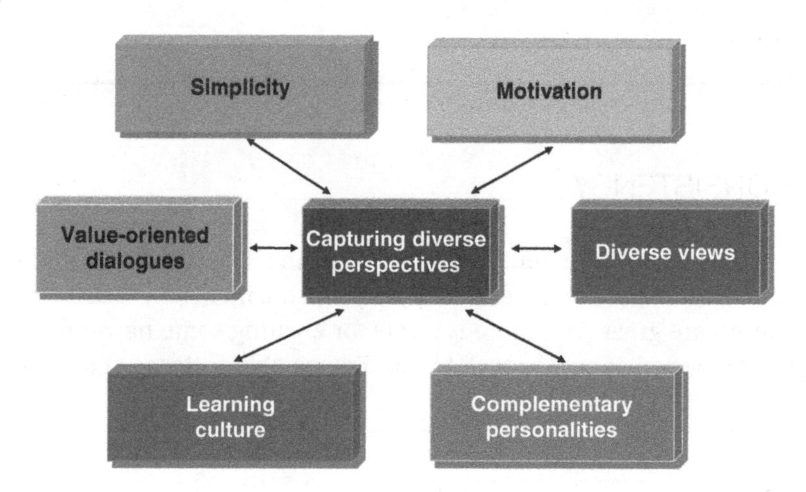

FIGURE 14.5 Diverse Perspectives

The So What:

- Culture maps, understanding global regions, collaborating environments across teams
- Getting the right support, working closely with benefits sponsors, right voices are brought to the mix
- Encouraging and proper emphasis are exercised, and the servant leadership's importance to setting the stage of where the behaviors shift could help us achieve the change success

Review Questions

Parentheses ( ) are used for Multiple Choice, when one answer is correct. Brackets [ ] are used for Multiple Answer, when many answers are correct.

_____ is an example of what servant leadership could contribute to facilitating the voice of the program team.

( ) Lead from the top
( ) Rolling up the sleeves to show humility and drive change
( ) Management team first
( ) Outputs focused interactions.

(continued)

(*continued*)

Which of the following is a direct benefit to understanding culture maps for a program team? Choose all that apply

[ ] Developing solutions to serve vocal team members.
[ ] Capitalizing on the strengths of various cultures.
[ ] Co-creating better change roadmaps.
[ ] Using the dominant culture views.

Which is of the following are examples of how benefit sponsors help program managers in diverse views?

( ) Increasing the program team's size.
( ) Focus on financial results.
( ) Practice value-oriented dialogues.
( ) Using detailed roadmaps.

14.5 DIVERSITY CONSISTENCY

Such a powerful topic as program management is designed to bring different teams and teams of teams together and integrate their ideas to the mix. The richness of diversity and the inclusion of different ideas in the mix of the program plans and execution are great. Yet, there is a role for creating some harmony in the direction and path forward, while achieving some level of consistency to build repeatable patterns of success.

Key Learnings

- Leadership qualities to support adapting
- Value focus as in the case of adaptable roadmaps
- The cross-cutting programs that inspire change

14.5.1 Ronald Heifetz on Leadership

According to Heifetz: "***Adaptive leadership is the practice of mobilizing people to tackle tough challenges and thrive***" – brings together many of the points about resiliency and the co-created adaptable roadmaps.

Why risk is key ingredient to leadership consistency?

- Risk ownership = ***governance excellence*** (risk-based)
- Need to develop the stomach to deal with program change uncertainty (key quality to develop in programs work)
- Build the muscles of continual adapting and innovating (a model for future leadership, creating the foundation for injecting different views and purring the right value on these views)

14.5.2 Consistent Program Change Maker

Great reminders and summary points for the role of an effective and consistent Program Change Maker:

- Holistic views (getting on the balcony rather than being on the dance floor all the time)
- Joint ownership and voices (as in giving the work back to people)
- Dealing with D-VUCAD (extended version of VUCA, adding disruptions in the environment, and the diversity which brings the strength to the quality of the roadmaps we create for achieving change)

14.5.3 Experimenting and Failing

Most talked about principle across learning organizations

- Experimenting is a vulnerable process
- This ownership to program value is anchored in learning fast
- Diversity potential increases with collaboration (need to see and educate others on that potential to invest properly in this key principle)

14.5.4 Active Listening Tactics

As in the case of the soccer team picture here

Source: garten-gg / 4865 images.

What does the future of active listening look like?

- Dynamic and holistic in nature
- Consistent use of pause and safe spaces (power of pause opens the dialogues)
- Across all layers of the program team
- Value-Centered (keeping an eye on the goal as in the case of the soccer team, open my eyes to the diversity available in the program team)

14.5.5 Start, Stop, and Continue

Source: BubbleJuice / 34 images.

- Creates an adaptive muscle (develop our own worksheets to help us implement and adapt)
- Enhances the opportunity to listen to multiple views and rethink course of action (comparing notes with colleagues and others)
- The manufacturing company PMO failure story in practice (Sometime learning from a failing story to find better ways to connecting to stakeholders, having the right focused roadmap, making sure we use the learning worksheets to be more aware of diversity and to turn the tide towards value)
- A clear diversity differentiator

The So What:

- D-VUCAD adds disruptions diversity elements which brings more valuable adaptability to program roadmaps
- Consistent use of pause, safe spaces, and adaptive models of leadership allow for the stretching of program teams

Review Questions

Parentheses ( ) are used for Multiple Choice, when one answer is correct. Brackets [ ] are used for Multiple Answer, when many answers are correct.

_____ illustrates that program leaders are capable to adapt.

( ) The program leader should stick to the tried and true ways.
( ) Mobilizing the team to tackle challenges.
( ) Governance is a board role only.
( ) Team members should avoid risks.

Which of the following are ingredients of effective program experimenting? Choose all that apply

[ ] Only in the hands of the program manager.
[ ] Taking extra time to learn.
[ ] Program managers should ensure there is clear value ownership.
[ ] Being open to diverse views.

Which is of the following is a clear sign of the future of active listening?

( ) All decisions are deliverables centered.
( ) Slow decision-making process.
( ) Consistent use of pause and safe spaces.
( ) Executives make the best decisions.

SECTION V

THE PATH FORWARD

Just like in projects and programs, the closure milestone is a critical milestone, the last section of this book. Programs have a holistic nature like this journey. And if we have gotten into the core focus of the book, which is the programs' ability to enable the processes, people, and supporting technology, to drive change and achieve value, we would know that there is value and learning to be achieved all the way to the end of a program and extends further beyond.

Programs and projects are like labs for experimenting, learning, ways of working, and trying different ingredients to test the boundaries of the potential of the team, the quality of the program plans, and the ability of the organization to deliver a highly connected set of outcomes to the customers/stakeholders of the program/project.

Understanding the programs, our role in the programs, build on the core focus of this book, the focus on change, the right outcomes, the enablers for change like the culture, change scientists, and the criticality of adaptable roadmaps and their focus on cocreation and diversity.

Benefiting from Erin Meyer's empathy views, Simon Sinek's program leadership unique attributes, adaptable framework models, and many of the other principles examples all hinge on appreciating the diversity of expertise and points of views of the program team, and integrating that to help achieve sustained success.

STRATEGIC OPPORTUNITIES FOR PROGRAM MANAGEMENT

One of the critical impacts of programs and program management is their impact on business models and how the business models impact the ways by which programs are delivered. This is increasingly a fact of life in the future project and program economy.

Program Management and Business Models[1]

In some companies, programs are initiated to produce products and services that have the potential of creating long-term business value. The way the organization delivers and captures the desired value for economic gain is through the company's business model, which interacts within the marketplace to take advantage of a commercial opportunity. Simply stated programs and projects create value opportunities and business models deliver and capture the value of the programs generally in economic terms. Business model development focuses on how a

[1]Material in this section has been provided by Harold Kerzner and adapted from Harold Kerzner (2023)

Program Management: Going Beyond Project Management to Enable Value-Driven Change, First Edition. Al Zeitoun.

firm intends to maximize the usage of its resources and core competencies. Today's program managers must have an understanding of business models.

In theory and practice, the term *business model* is used for a broad range of informal and formal descriptions to represent core aspects of a business, including purpose, business process, target customers, offerings, strategies, infrastructure, organizational structures, sourcing, trading practices, and operational processes and policies including culture. The process of business model construction is part of business strategy and requires participation of program managers. At the start of a program, the program manager must determine whether the program will be executed within the firm's existing business model or whether a new business model must be created as the program evolves. The need for a new business model can impact the downstream decisions made by program managers.

Business models are used to describe and classify businesses, especially in an entrepreneurial setting, but they are also used by managers inside companies to explore possibilities for future development and to increase competitiveness. Business model development is most frequently a trial-and-error process. Business models must be reassessed frequently by the program managers due to rapidly changing markets and a high degree of business uncertainty.

Traditional program management is often viewed as a structured environment with frequently well-defined requirements at program initiation and constraints on time, cost, and scope. Innovation for new products and services within the program is often considered as work that is free flow, without boundaries, and with very little pressure from the triple constraints of time, cost, and scope. However, this does not mean that innovation within the program will run forever.

For the creation of a new business model or for changes needed to an existing business model as the result of a program's activities, program managers may find themselves in a completely unfamiliar environment. As examples:

- There may be no business model statement of work
- Program managers are brought on board before a business model business case is developed (assume one is developed) and must interface with the customers and users in the marketplace in the meanwhile
- Program managers may need to participate in market research for the business model
- Program managers may need to perform a great many experiments before finalizing the business model
- Program managers may have to build and test numerous products and/or services prototypes before finalizing the business model
- Program managers must understand the organization's core competencies concerning resources and capabilities, and how they fit into the design of a business model
- Program managers may need a completely diverse set of tools for business model development

The program environment gets even worse when companies try to use inappropriate program management practices for business processes such as opting for simple business model changes where you have the greatest degree of risk and uncertainty, where traditional risk management planning will not work and where a great deal of flexibility is needed for decision-making. Different approaches, many requiring a high level of flexibility, will be differentiated by the level of technology, the amount of product changes, and whether the impact is expected to disrupt the markets.

From Program Manager to Business Model Designer

Based upon the newness and size of the program, new business model design may require that the program manager assume the lead role as the designer of the model. According to Van Der Pilj et al. (2016, p. 9):

> *Design is fundamentally about enhancing the way you look at the world. It's a learnable, repeatable, disciplined process that everyone can use to create unique and qualified value. Design is not about throwing away the processes and tools you have. In fact, quite the opposite is true. Just as design has enabled countless upstarts to create new business models and markets, design will also help you decide when to use what tools in order to learn something new, persuade others to take a different course, and at the end of the day, make better (business) decisions.*

Most of all, design is about creating the conditions by which businesses thrive, grow, and evolve in the face of uncertainty and change. As such, better businesses are ones that approach problems in a new, systematic way, focusing more on doing rather than on planning and prediction. Better businesses marry design and strategy to harness opportunity in order to drive growth and change in a world that is uncertain and unpredictable.

The skills needed for the design effort requires imagination and experimentation and testing of a variety of models to identify the best approach. As stated by Kaplan (2012, pp. 142–143):

As a Business Model Designer, you will design for change by

- Leading and contributing to ethnographic fieldwork to generate powerful customer experience observations and insights
- Leading and contributing to design teams through analysis and synthesis of ethnographic work to distill the most important insights leading to transformational business model concepts
- Developing testable business model concepts and prototypes
- Leading and contributing to real-world business model experiments
- Creating and implementing frameworks to measure results and impact of business model experiments
- Crafting and executing compelling multimedia stories that help stakeholders understand and connect with the work
- Capturing and packaging learning from business model experiments to inform other efforts and ensure maximum leverage

Business Models and Business Value

Business models are often treated as value networks that convert the outcome of programs to business value. Program managers must recognize that innovations can take place in processes within the business model to bring added value to the programs. Included in the business model can be the relationships/partnerships and contractual agreements with the suppliers and buyers. The relationship with the suppliers and the buyers can be highly advantageous and bring long-term business value to all parties if each party has similar business value drivers such that the business relationship is mutually beneficial to all. In such situations, it becomes important when evaluating potential business partners to make sure that there are similar interpretations of business value so that the business models of each party are complementary. Effective design and enhancements to a business model can give a company a sustainable competitive advantage.

Value is like beauty; it is in the eyes of the beholder. Highly diversified companies may need different business models for whom they are creating value. Typically, customer segments can be classified as mass markets, niche markets, segmented markets, and diversified markets. Each market segment may require a different business model, and each company within the segment may have a different interpretation of the value being provided through your business models. Program managers must understand these differences.

Business Model Characteristics

Every company has a unique business model based on their core competencies and technologies they use. However, there are some generic commonalities between business models that fit most programs. Business model design, as described by Osterwalder and Pigneur (2010), includes the modeling and description of a company's:

- Value propositions (What value and benefits will the customers receive from the products and services?)
- Target customer segments (What desires and expectations do the customers have when purchasing your products and services?)
- Distribution channels (What are the ways that a company keeps in touch with its clients?)
- Customer relationships (What are the different links that a company can use to maintain relations with various market segments?)

- Value configurations (How have we arranged our resources and activities to bring value to our clients?)
- Core capabilities (What are our core competencies that support our business model?)
- Partner network (What are our partnership agreements with other companies that assist us in providing commercialized value?)
- Cost structure (How have we leveraged our financial investments in the business model to provide commercialized value?)
- Revenue generation model (What are the ways in which the business model supports revenue flows?)

Osterwalder refers to these nine items as the business model canvas, which outlines your business model in the shape of a story to describe how you create, deliver, and capture value. Program managers should not assume that the program team members fully understand the company's business model. By preparing a business model canvas, it is easier to understand the business model attributes and the impact on the programs.

If you have direct competitors that appear to have a good business model, it may be beneficial to create a business model canvas for their business and see their strengths, weaknesses, opportunities, and threats. This may provide you with ideas about your new business model.

Strategic Partnerships

It is unrealistic for a program manager to expect companies to always own all the resources needed to remain competitive. Companies view strategic partnerships as a way of building competitive business models. Program managers must understand the importance of building strategic partnerships. Osterwalder and Pigneur (2010, p. 38) identify four types of partnerships:

1. Strategic alliances between noncompetitors
2. *Cooperation*: strategic partnerships between competitors
3. Joint ventures to develop new businesses
4. Buyer–supplier relationships to ensure reliable supplies

There are several benefits in these relationships. The partners either employ or have access to resources that are critical to your business model. The partners can help reduce risk and uncertainty as well assisting you in identifying and responding to threats. You may be able to access new market segments through the business models of your strategic partners.

In traditional project management activities, project and program managers are accustomed to working with suppliers and partners for the procurement of materials and components. Quite often, these activities are handled through a contract administrator. With program management activities, the burden of responsibility falls on the program manager who must interface with the partner's business personnel and make business rather than technical decisions. This mandates an understanding of your own business model as well as those of your strategic partners.

Business Intelligence

Business intelligence is information that a company needs to make the best business decisions based on facts and evidence rather than just guesses. This information may be stored in a database or data warehouse. Business intelligence technologies provide historical, current, and predictive views of business operations. Common functions of business intelligence technologies include reporting, online analytical processing, analytics, data mining, process mining, complex event processing, business performance management, benchmarking, text mining, predictive analytics, and prescriptive analytics.

Business intelligence can be used by enterprises to support a wide range of business decisions from operational to strategic. Basic operating decisions include product positioning or pricing. Strategic business decisions involve priorities, goals, and directions at the broadest level.

Business intelligence tools allow organizations to gain insight into new markets, to assess demand and suitability of products and services for different market segments and to gauge the impact of marketing efforts. This can result in a competitive market advantage and long-term stability. Business intelligence can identify the following:

- Strategic opportunities for new products and services requiring innovation
- Strategic opportunities for improvements in your company's business model
- Strengths and weaknesses in the business models of your competitors
- Strengths and weaknesses in your company's business model and where competitors may attack

Historically, many program managers had a poor understanding of how the business functions. Business-related decisions were usually placed in the hands of the project sponsors or governance committees. Today, this has changed. Some companies believe that program managers of the future will possess significantly more business knowledge.

Business Model Enhancements

Companies invest a great deal of time and effort for a program's new products and services. Unfortunately, considerably less effort is applied to the firm's business model until a threat appears. Most firms are relatively slow in identifying threats and even when they see them, there is often a complacency attitude and a slow response time in deciding whether enhancements are needed to the existing business model or if a totally new model is needed. The competition between most companies today is between business models, not with the products and services provided (Gassmann et al. 2013). Products and services can be duplicated, whereas business models are usually unique to a firm because they tap into the firm's specialized strengths and competencies.

Several factors can act as triggers indicating that the business model processes must be reevaluated:

- Changes in consumers' needs for your products and services
- New competitors entering the marketplace
- New suppliers entering the marketplace
- Changing relationships with your strategic partners in the business model
- Significant changes in your firm's core competencies
- Significant changes in the assumptions made about the enterprise environmental factors

Lüttgens and Diener (2016) looked at the threats to a business model using Porter's five forces that describe the competitive forces within an industry. Porter defined the five factors as the bargaining power of buyers, bargaining power of suppliers, competitive rivalry, threats of new entrants, and threats of substitutes. Changes in these five forces can be used as early warning signs or triggers that there are threats to the business model.

The purpose of a business model is to create business value in a profitable manner. Therefore, enhancements or innovations must focus on the processes associated with the five dimensions of value (Baden-Fuller and Morgan 2010; Beinhocker 2007; Abdelkafi et al. 2013):

1. Value proposition
2. Value creation
3. Value communication
4. Value in distribution channels
5. Value capture

The result of process innovations can be either added value or cost reductions.

Core competencies are the building blocks for many forms of business value and serve as the basis for the firm's competitive posture. The core competencies are the combination of the firm's resources, knowledge, and skills that create core products for the end users. These core products contribute to the competitiveness of the firm by accessing a wide variety of markets and making it difficult for competitors to imitate your products and the accompanying value perceived by the customers. Management must look for ways of improving the core competencies to create new products and new markets.

Business Models and Strategic Alliances

Companies frequently develop close relationships and strategic alliances with companies that are part of the supply chain during innovation activities such as new product development. The result is usually a win–win situation for all parties with the expectation that these relationships will continue for some time into the future. If a small change to a product is necessary, the impact on the relationships is usually minor. But if the company is focusing on a new product that will result in a new business model, the relationships with the supply chain suppliers can be destroyed if the suppliers believe that the new business model will provide them with fewer benefits than before. As such, the suppliers may not wish to provide support for the new product or new business model development if they believe it is not in their best interest.

When a company seeks to disrupt a market by introducing a new product, or responding to disruptions caused by a competitor, the company may need to replace its core competencies with new ones and likewise develop other supply chain relationships. If the company's core competencies are rigid and the company depends heavily on the core competencies of their suppliers, the company may wish to reconsider innovation activities that could result in a new business model that would destroy existing relationships. Characteristics of strategic alliances, adapted from Spekman et al. (2000) include:

- *Trust*: Based on established norms, values, past experiences, and reputation
- *Commitment*: Sharing critical and proprietary information over the long term
- *Interdependence*: Partnership cooperation and dependence on one another for the long term
- *Cultural compatibility*: Trying to align cultures that support a close working relationship
- *Planning and coordination*: A joint effort focusing on the future of the relationship

Most business model designs are impacted by partnerships and alliances in the supply chain. Therefore, any decision made by the program manager to alter a business model without considering the impact on supply chain relationships might be a mistake.

Identifying Business Model Threats

There is an old adage that experienced project and program managers often follow, hope for the best but plan for the worst. The failure of a business model can have catastrophic consequences for a company. When developing a business model, program managers must always ask themselves, what can go wrong? Simply stated, the company must perform risk management and determine the threats to their business model. Sometimes, the threats are not apparent during the development stage. Periodically, companies must therefore reassess all possible threats in time to react appropriately.

There is no standard approach for assessing business model threats. One way to assess threats is by using Porter's five forces, namely (i) entry barriers, (ii) exit barriers, (iii) bargaining power of suppliers, (iv) bargaining power of buyers, and (v) substitute products. As an example, some companies are highly dependent on suppliers for components and materials. We should then assess the threat of what could happen if the supplier is late, refuses to work with us, want to increase their prices, and other such situations.

Business Model Failure

Senior management must have a vision on how they want the company to compete, now and in the future. This information must be provided to the program managers responsible for creating or enhancing a business model. Most executives understand that they must compete through business models rather than just products and services, but they lack an understanding of how to do it. The result is usually a doomed business model.

Kaplan (2012, pp. 40–49) identifies 10 reasons and attitudes that cause companies to fail at business model innovation:

1. CEOs don't really want a new business model.
2. Business model innovation will be the next CEO's problem.
3. Product is king – nothing else matters.
4. Information technology is only about keeping the trains moving and lowering costs.
5. Cannibalization is off the table.
6. Nowhere near enough connecting with unusual suspects.
7. Line executives hold your pay card.
8. Great idea; what's the ROI?
9. They shoot business model innovators, don't they?
10. You want to experiment in the real world; are you crazy?

There are seven common mistakes made by senior management:

1. Not realizing that good business models lead to a sustainable competitive advantage
2. Not viewing your business model through the eyes of your customers
3. Building a business model in isolation without considering how your competitors might react and potential threats
4. Building a business model in isolation without considering how you will interact with the business models of your competitors
5. Refusing to build a business model that is new to your firm regardless of whether the competitors have a similar model
6. Believing that your current business model does need to undergo continuous improvement efforts
7. Not understanding that business models are more than just products and services and must include such items as sales and marketing activities, procurement practices, strategic partnerships, opportunities for vertical integration, and compensation practices

The following case studies[2] give you an opportunity to connect the dots across the sections of this book and allows you to stretch yourself. As you tackle the questions for each case, please go back and review every takeaway, every skill, every story, concept you noted as we progressed on this book's journey and internalize that reflection toward achieving consisting value out of your programs and your work.

CASE STUDY: NORA'S DILEMMA

Nora was now having second thoughts about whether she made the right career choice wanting a future as a program or project manager. She was also unsure if she should continue working for Dexter Aerospace Corporation.

[2]The case studies, "The Blue Spider Project," "McRoy Aerospace," "The Team Meeting," "The Prima Donna," "Zane Corporation," and "The Management Control Freak" have been taken and repurposed from Kerzner (2022)

Six months ago, Nora completed her master's degree in business administration with a minor in project management. She wanted a career in project management. She hired into Dexter Corporation as a project manager after graduation. Dexter had numerous government contracts for exploratory satellites, and these were the types of projects that intrigued Nora.

Nora's first assignment was to work with Dexter's sales team to respond to a government request for proposal (RFP) for a new series of satellites to explore the properties of the sun. The RFP identified the requirements and technical specifications that had to be met for the development and testing of three prototypes. A follow-on contract would then be awarded to the winner of the RFP for the manufacturing of several satellites.

A meeting was held with the lead salesperson, Nora, and the engineering and manufacturing personnel that would be preparing the technical portions of the proposal.

The lead salesperson then made the following comments:

> Senior management considers this potential contract as very important for Dexter's future and we must submit a winning bid. When estimating the work needed by your organizations and the accompanying costs, base your estimates upon the absolute minimum work Dexter must do to satisfy the requirements. We want to submit the lowest cost bid.
>
> Then look for loopholes and omissions in the requirements stated in the RFP and prepare a list of all the scope changes and accompanying costs we could possibly generate after contract go-ahead. There are always things that the government did not consider, and must be accomplished, but let's not tell them about these things they neglected to consider other than through the scope changes we can generate after contract award.
>
> Also, Dexter has never tested any products in the temperature range identified in the technical specifications. To get a step up on our competitors, let's include some wording asserting that we have done a little bit of testing in the temperature range requested and the results were promising. This should help us win the contract.

Nora could not believe what she had heard. The lead salesperson's comments seemed to violate what Nora learned in college about business morality and ethics and seemed to contradict PMI's code of professional responsibility. After the meeting was adjourned, Nora met with the lead salesperson and asked:

> Why aren't we 100% honest with the government about all of the work that needs to be done to achieve project success and fulfill the requirements?

The salesperson responded:

> The goal is to win the initial contract at all costs. It may look like we are "intentionally" lying to the customer, but we simply consider this as our initial interpretation of the requirements.
>
> Sometimes, we might even bid the initial contract at a significant loss just to win it. Then, we push through the very profitable scope changes that most often generate significantly more profit than the initial contract. This is a 'way of life' in our industry, and you'll need to get used to it. It's probably the scope changes that will be paying your salary rather than the initial contract.

Nora then asked:

> Doesn't the government know this is happening?

The salesperson then replied:

> Yes, I am sure they know this is happening. Once the initial contract is awarded, the government as well as other customers we have would rather go along with the approval and funding of many scope changes than

to repeat the acquisition process and go out for competitive bidding again looking for new suppliers. On some of our contracts, the people that approve the initial contract and follow-on scope changes are military personnel that have just a two- or three-year tour of duty in this assignment and then get transferred to another assignment elsewhere. Whoever replaces them may then have to go before Congress or other approval agencies and explain the reasons for the cost overruns. This is how many of the aerospace and defense industry firms operate. Cost overruns are a way of life.

Nora looked at the salesperson and then said:

I have one more question. If you know Dexter has never done any experimentation in the temperature range requested by the client, why should we lie to the customer?

The salesperson replied:

We are not lying. We consider this as just misinterpretation of the facts, or just an error in wording someone made. I am sure that somewhere in the labs are test results that we could "recreate" confirming our wording in the proposal.

Nora read over the entire proposal prior to submittal to the government. As expected, included in the proposal was sort of vague wording that Dexter had some previous experience in testing products in the customer's temperature specification range. The proposal was sent off to the customer and Dexter expected to hear whether they won the contract within 30 days.

In less than two weeks, Nora was asked to attend an emergency meeting with the lead salesperson on the proposal. The salesperson looked at Nora and said:

The government wants to visit our company quickly as see the test results we stated in our proposal about the testing we did within the specification's temperature range. How do you think we should handle their request since it may have a serious impact on who is awarded the contract? Think about your answer and let me know tomorrow.

Nora now had second thoughts about whether she should be a project manager at Dexter or anywhere else? Could this happen elsewhere she thought?

The So What

- Think about how you would handle this type of unprofessional situation if encountered in your work journey
- In a better situation than highlighted in this case, consider what role Enterprise Risk Management (ERM) could play in being better prepared

CASE STUDY: THE BLUE SPIDER PROJECT

"This is impossible! Just totally impossible! Ten months ago, I was sitting on top of the world. Upper-level management considered me one of the best, if not the best, engineer in the plant. Now look at me! I have bags under my eyes, I haven't slept soundly in the last six months, and here I am, cleaning out my desk. I'm sure glad they gave me back my old job in engineering. I guess I could have saved myself a lot of grief and aggravation had I not accepted the promotion to project manager."

History

Gary Anderson had accepted a position with Parks Corporation right out of college. With a PhD in mechanical engineering, Gary was ready to solve the world's most traumatic problems. At first, Parks Corporation offered Gary little opportunity to do the pure research that he eagerly wanted to undertake. However, things soon changed. Parks grew into a major electronics and structural design corporation during the big boom of the late 1950s and early 1960s when Department of Defense (DoD) contracts were plentiful.

Parks Corporation grew from a handful of engineers to a major DoD contractor, employing some 6500 people. During the recession of the late 1960s, money became scarce and major layoffs resulted in lowering the employment level to 2200 employees. At that time, Parks decided to get out of the R&D business and compete as a low-cost production facility while maintaining an engineering organization solely to support production requirements.

After attempts at virtually every project management organizational structure, Parks Corporation selected the matrix form. Each project had a program manager who reported to the director of program management. Each project also maintained an assistant project manager – normally a project engineer – who reported directly to the project manager and indirectly to the director of engineering. The program manager spent most of his time worrying about cost and time, whereas the assistant program manager worried more about technical performance.

With the poor job market for engineers, Gary and his coworkers began taking coursework toward MBA degrees in case the job market deteriorated further. In 1995, with the upturn in DoD spending, Parks had to change its corporate strategy. Parks had spent the last seven years bidding on the production phase of large programs. Now, however, with the new evaluation criteria set forth for contract awards, those companies winning the R&D and qualification phases had a definite edge on being awarded the production contract. The production contract was where the big profits could be found. In keeping with this new strategy, Parks began to beef up its R&D engineering staff. By 1998, Parks had increased in size to 2700 employees. The increase was mostly in engineering. Experienced R&D personnel were difficult to find for the salaries that Parks was offering. Parks was, however, able to lure some employees away from the competitors, but relied mostly upon the younger, inexperienced engineers fresh out of college.

With the adoption of this corporate strategy, Parks Corporation administered a new wage and salary program that included job upgrading. Gary was promoted to senior scientist, responsible for all R&D activities performed in the mechanical engineering department. Gary had distinguished himself as an outstanding production engineer during the past several years, and management felt that his contribution could be extended to R&D as well.

In January 1998, Parks Corporation decided to compete for Phase I of the Blue Spider Project, an R&D effort that, if successful, could lead into a \$500 million program spread out over 20 years. The Blue Spider Project was an attempt to improve the structural capabilities of the Spartan missile, a short-range tactical missile used by the Army. The Spartan missile was exhibiting fatigue failure after six years in the field. This was three years less than what the original design specifications called for. The Army wanted new materials that could result in a longer life for the Spartan missile.

Lord Industries was the prime contractor for the Army's Spartan Program. Parks Corporation would be a subcontractor to Lord if they could successfully bid and win the project. The criteria for subcontractor selection were based not only on low bid but also on technical expertise as well as management performance on other projects. Park's management felt that it had a distinct advantage over most of the other competitors because they had successfully worked on other projects for Lord Industries.

The Blue Spider Project Kickoff

On November 3, 1997, Henry Gable, the director of engineering, called Gary Anderson into his office.

Henry Gable: "Gary, I've just been notified through the grapevine that Lord will be issuing the RFP for the Blue Spider Project by the end of this month, with a 30-day response period. I've been waiting a long time for a project like this to come along so that I can experiment with some new ideas that I have. This project is going to be my baby all the way! I want you to head up the proposal team. I think it must be an engineer. I'll make sure that you get a good proposal manger to help you. If we start working now, we can get close to two months of research in before proposal submittal. That will give us a one-month's edge on our competitors."

Gary was pleased to be involved in such an effort. He had absolutely no trouble in getting functional support for the R&D effort necessary to put together a technical proposal. All of the functional managers continually remarked to Gary, "This must be a biggy. The director of engineering has thrown all of his support behind you."

On December 2, the RFP was received. The only trouble area that Gary could see was that the technical specifications stated that all components must be able to operate normally and successfully through a temperature range of −65 to 145 °F. Current testing indicated the Parks Corporation's design would not function above 130 °F. An intensive R&D effort was conducted over the next three weeks. Everywhere Gary looked, it appeared that the entire organization was working on his technical proposal.

A week before the final proposal was to be submitted, Gary and Henry Gable met to develop a company position concerning the inability of the preliminary design material to be operated above 130 °F.

Gary Anderson: "Henry, I don't think it is going to be possible to meet specification requirements unless we change our design material or incorporate new materials. Everything I've tried indicates we're in trouble."

Gable: "We're in trouble only if the customer knows about it. Let the proposal state that we expect our design to be operative up to 155 °F. That'll please the customer."

Anderson: "That seems unethical to me. Why don't we just tell them the truth?"

Gable: "The truth doesn't always win proposals. I picked you to head up this effort because I thought that you'd understand. I could have just as easily selected one of our many moral project managers. I'm considering you for program manager after we win the program. If you're going to pull this conscientious crap on me like the other project managers do, I'll find someone else. Look at it this way; later we can convince the customer to change the specifications. After all, we'll be so far downstream that he'll have no choice."

After two solid months of 16-hour days for Gary, the proposal was submitted. On February 10, 1998, Lord Industries announced that Parks Corporation would be awarded the Blue Spider Project. The contract called for a 10-month effort, negotiated at $2.2 million at a firm-fixed price.

Selecting the Project Manager

Following contract award, Henry Gable called Gary in for a conference.

Gable: "Congratulations, Gary! You did a fine job. The Blue Spider Project has great potential for ongoing business over the next 10 years, provided that we perform well during the R&D phase. Obviously, you're the most qualified person in the plant to head up the project. How would you feel about a transfer to program management?"

Anderson: "I think it would be a real challenge. I could make maximum use of the MBA degree I earned last year. I've always wanted to be in program management."

Gable: "Having several masters' degrees, or even doctorates for that matter, does not guarantee that you'll be a successful project manager. There are three requirements for effective program management: You must be able to communicate both in writing and orally; you must know how to motivate people; and you must be willing to give up your car pool. The last one is extremely important in that program managers must be totally committed and dedicated to the program, regardless of how much time is involved.

"But this is not the reason why I asked you to come here. Going from project engineer to program management is a big step. There are only two places you can go from program management – up the organization or out the door. I know of very, very few engineers who failed in program management and were permitted to return."

Anderson: "Why is that? If I'm considered to be the best engineer in the plant, why can't I return to engineering?"

The Work Begins

Gable: "Program management is a world of its own. It has its own formal and informal organizational ties. Program managers are outsiders. You'll find out. You might not be able to keep the strong personal ties you now have with your fellow employees. You'll have to force even your best friends to comply with your standards. Program managers can go from program to program, but functional departments remain intact.

"I'm telling you all this for a reason. We've worked well together the past several years. But if I sign the release so that you can work for Grey in program management, you'll be on your own, like hiring into a new company. I've already signed the release. You still have some time to think about it."

Anderson: "One thing I don't understand. With all of the good program managers we have here, why am I given this opportunity?"

Gable: "Almost all of our program managers are over 45 years old. This resulted from our massive layoffs several years ago when we were forced to lay off the younger, inexperienced program managers. You were selected because of your age and because all of our other program managers have worked only on production-type programs. We need someone at the reins who knows R&D. Your counterpart at Lord Industries will be an R&D type. You have to fight fire with fire. "I have an ulterior reason for wanting you to accept this position. Because of the division of authority between program management and project engineering, I need someone in program management whom I can communicate with concerning R&D work. The program managers we have now are interested only in time and cost. We need a manager who will bend over backwards to get performance also. I think you're that man. You know the commitment we made to Lord when we submitted that proposal. You have to try to achieve that. Remember, this program is my baby. You'll get all the support you need. I'm tied up on another project now. But when it's over, I'll be following your work like a hawk. We'll have to get together occasionally and discuss new techniques.

"Take a day or two to think it over. If you want the position, make an appointment to see Elliot Grey, the director of program management. He'll give you the same speech I did. I'll assign Paul Evans to you as chief project engineer. He's a seasoned veteran and you should have no trouble working with him. He'll give you good advice. He's a good man."

Gary accepted the new challenge. His first major hurdle occurred in staffing the project. The top priority given to him to bid the program did not follow through for staffing. The survival of Parks Corporation depended on the profits received from the production programs. In keeping with this philosophy, Gary found that engineering managers (even his former boss) were reluctant to give up their key people to the Blue Spider Program. However, with a little support from Henry Gable, Gary formed an adequate staff for the program.

Right from the start Gary was worried that the test matrix called out in the technical volume of the proposal would not produce results that could satisfy specifications. Gary had 90 days after go-ahead during which to identify the raw materials that could satisfy specification requirements. Gary and Paul Evans held a meeting to map out their strategy for the first few months.

Anderson: "Well, Paul, we're starting out with our backs against the wall on this one. Any recommendations?"

Paul Evans: "I also have my doubts about the validity of this test matrix. Fortunately, I've been through this before. Gable thinks this is his project and he'll sure as hell try to manipulate us. I have to report to him every morning at 7:30 a.m. with the raw data results of the previous day's testing. He wants to see it before you do. He also stated that he wants to meet with me alone.

"Lord will be the big problem. If the test matrix proves to be a failure, we're going to have to change the scope of effort. Remember, this is an FFP contract. If we change the scope of work and do additional work in the earlier phases of the program, then we should prepare a trade-off analysis to see what we can delete downstream so as to not overrun the budget."

Anderson: "I'm going to let the other project office personnel handle the administrating work. You and I are going to live in the research labs until we get some results. We'll let the other project office personnel run the weekly team meetings."

For the next three weeks, Gary and Paul spent virtually 12 hours per day, 7 days a week in the research and development lab. None of the results showed any promise. Gary kept trying to set up a meeting with Henry Gable but always found him unavailable.

During the fourth week, Gary, Paul, and the key functional department managers met to develop an alternate test matrix. The new test matrix looked good. Gary and his team worked frantically to develop a new workable schedule that would not have impact on the second milestone, which was to occur at the end of 180 days. The second milestone was the final acceptance of the raw materials and preparation of production runs of the raw materials to verify that there would be no scale-up differences between lab development and full-scale production.

Gary personally prepared all of the technical handouts for the interchange meeting. After all, he would be the one presenting all of the data. The technical interchange meeting was scheduled for two days. On the first day, Gary presented all of the data, including test results, and the new test matrix. The customer appeared displeased with the progress to date and decided to have its own inhouse caucus that evening to go over the material that was presented.

The following morning the customer stated its position: "First of all, Gary, we're quite pleased to have a project manager who has such a command of technology. That's good. But every time we've tried to contact you last month, you were unavailable or had to be paged in the research laboratories. You did an acceptable job presenting the technical data, but the administrative data was presented by your project office personnel. We, at Lord, do not think that you're maintaining the proper balance between your technical and administrative responsibilities. We prefer that you personally give the administrative data and your chief project engineer present the technical data.

"We did not receive any agenda. Our people like to know what will be discussed, and when. We also want a copy of all handouts to be presented at least three days in advance. We need time to scrutinize the data. You can't expect us to walk in here blind and make decisions after seeing the data for 10 minutes.

"To be frank, we feel that the data to date is totally unacceptable. If the data does not improve, we will have no choice but to issue a work stoppage order and look for a new contractor. The new test matrix looks good, especially since this is a firm-fixed-price contract. Your company will bear the burden of all costs for the additional work. A trade-off with later work may be possible, but this will depend on the results presented at the second design review meeting, 90 days from now.

"We have decided to establish a customer office at Parks to follow your work more closely. Our people feel that monthly meetings are insufficient during R&D activities. We would like our customer representative to have daily verbal meetings with you or your staff. He will then keep us posted. Obviously, we had expected to review much more experimental data than you have given us.

"Many of our top-quality engineers would like to talk directly to your engineering community, without having to continually waste time by having to go through the project office. We must insist on this last point.

Remember, your effort may be only $2.2 million, but our total package is $100 million. We have a lot more at stake than you people do. Our engineers do not like to get information that has been filtered by the project office. They want to help you.

"And last, don't forget that you people have a contractual requirement to prepare complete minutes for all interchange meetings. Send us the original for signature before going to publication."

Although Gary was unhappy with the first team meeting, especially with the requests made by Lord Industries, he felt that they had sufficient justification for their comments. Following the team meeting, Gary personally prepared the complete minutes. "This is absurd," thought Gary. "I've wasted almost one entire week doing nothing more than administrative paperwork." Why do we need such detailed minutes? Can't a rough summary suffice? Why is it that customers want everything documented? That's like an indication of fear. We've been completely cooperative with them. There has been no hostility between us. If we've gotten this much paperwork to do now, I hate to imagine what it will be like if we get into trouble."

A New Role

Gary completed and distributed the minutes to the customer as well as to all key team members.

For the next five weeks testing went according to plan, or at least Gary thought that it had. The results were still poor. Gary was so caught up in administrative paperwork that he hadn't found time to visit the research labs in over a month. On a Wednesday morning, Gary entered the lab to observe the morning testing. Upon arriving in the lab, Gary found Paul Evans, Henry Gable, and two technicians testing a new material, JXB-3.

Gable: "Gary, your problems will soon be over. This new material, JXB-3, will permit you to satisfy specification requirements. Paul and I have been testing it for two weeks. We wanted to let you know, but were afraid that if the word leaked out to the customer that we were spending their money for testing materials that were not called out in the program plan, they would probably go crazy and might cancel the contract. Look at these results. They're super!"

Anderson: "Am I supposed to be the one to tell the customer now? This could cause a big wave."

Gable: "There won't be any wave. Just tell them that we did it with our own IR&D funds. That'll please them because they'll think we're spending our own money to support their program."

Before presenting the information to Lord, Gary called a team meeting to present the new data to the project personnel. At the team meeting, one functional manager spoke out: "This is a hell of a way to run a program. I like to be kept informed about everything that's happening here at Parks. How can the project office expect to get support out of the functional departments if we're kept in the dark until the very last minute? My people have been working with the existing materials for the last two months and you're telling us that it was all for nothing. Now you're giving us a material that's so new that we have no information on it whatsoever. We're now going to have to play catch-up, and that's going to cost you plenty."

One week before the 180-day milestone meeting, Gary submitted the handout package to Lord Industries for preliminary review. An hour later the phone rang.

Customer: "We've just read your handout. Where did this new material come from? How come we were not informed that this work was going on? You know, of course, that our customer, the Army, will be at this meeting. How can we explain this to them? We're postponing the review meeting until all of our people have analyzed the data and are prepared to make a decision.

The Communications Breakdown

"The purpose of a review or interchange meeting is to exchange information when both parties have familiarity with the topic. Normally, we (Lord Industries) require almost weekly interchange meetings with our other customers because we don't trust them. We disregard this policy with Parks Corporation based on past working relationships. But with the new state of developments, you have forced us to revert to our previous position, since we now question Parks Corporation's integrity in communicating with us. At first, we believed this was due to an inexperienced program manager. Now, we're not sure."

Anderson: "I wonder if the real reason we have these interchange meetings isn't to show our people that Lord Industries doesn't trust us. You're creating a hell of a lot of work for us, you know."

Customer: "You people put yourself in this position. Now you have to live with it."

Two weeks later, Lord reluctantly agreed that the new material offered the greatest promise. Three weeks later the design review meeting was held. The Army was definitely not pleased with the prime contractor's recommendation to put a new, untested material into a multimillion-dollar effort.

During the week following the design review meeting, Gary planned to make the first verification mix in order to establish final specifications for selection of the raw materials. Unfortunately, the manufacturing plans were a week behind schedule, primarily because of Gary, since he had decided to reduce costs by accepting the responsibility for developing the bill of materials himself.

A meeting was called by Gary to consider rescheduling of the mix.

Anderson: "As you know we're about a week to 10 days behind schedule. We'll have to reschedule the verification mix for late next week."

Production manager: "Our resources are committed until a month from now. You can't expect to simply call a meeting and have everything reshuffled for the Blue Spider Program. We should have been notified earlier. Engineering has the responsibility for preparing the bill of materials. Why aren't they ready?"

Engineering integration: "We were never asked to prepare the bill of materials. But I'm sure that we could get it out if we work our people overtime for the next two days."

Anderson: "When can we remake the mix?"

Production manager: "We have to redo at least 500 sheets of paper every time we reschedule mixes. Not only that, we have to reschedule people on all three shifts. If we are to reschedule your mix, it will have to be performed on overtime. That's going to increase your costs. If that's agreeable with you, we'll try it. But this will be the first and last time that production will bail you out. There are procedures that have to be followed."

Testing engineer: "I've been coming to these meetings since we kicked off this program. I think I speak for the entire engineering division when I say that the role that the director of engineering is playing in this program is suppressing individuality among our highly competent personnel. In new projects, especially those involving R&D, our people are not apt to stick their necks out. Now our people are becoming ostriches. If they're impeded from contributing, even in their own slight way, then you'll probably lose them before the project gets completed. Right now, I feel that I'm wasting my time here. All I need are minutes of the team meetings and I'll be happy. Then I won't have to come to these pretend meetings anymore."

The purpose of the verification mix was to make a full-scale production run of the material to verify that there would be no material property changes in scale up from the small mixes made in the R&D laboratories. After testing, it became obvious that the wrong lots of raw materials were used in the production verification mix.

A meeting was called by Lord Industries for an explanation of why the mistake had occurred and what the alternatives were.

Lord: "Why did the problem occur?"

Anderson: "Well, we had a problem with the bill of materials. The result was that the mix had to be made on overtime. And when you work people on overtime, you have to be willing to accept mistakes as being a way of life. The energy cycles of our people are slow during the overtime hours."

Lord: "The ultimate responsibility has to be with you, the program manager. We, at Lord, think that you're spending too much time doing and not enough time managing. As the prime contractor, we have a hell of a lot more at stake than you do. From now on we want documented weekly technical interchange meetings and closer interaction by our quality control section with yours."

Anderson: "These additional team meetings are going to tie up our key people. I can't spare people to prepare handouts for weekly meetings with your people."

Lord: "Team meetings are a management responsibility. If Parks does not want the Blue Spider Program, I'm sure we can find another subcontractor. All you (Gary) have to do is give up taking the material vendors to lunch and you'll have plenty of time for handout preparation."

Gary left the meeting feeling as though he had just gotten raked over the coals. For the next 2 months, Gary worked 16 hours a day, almost every day. Gary did not want to burden his staff with the responsibility of the handouts, so he began preparing them himself. He could have hired additional staff, but with such a tight budget, and having to remake verification mix, cost overruns appeared inevitable.

As the end of the seventh month approached, Gary was feeling pressure from within Parks Corporation. The decision-making process appeared to be slowing down, and Gary found it more and more difficult to motivate his people. In fact, the grapevine was referring to the Blue Spider Project as a loser, and some of his key people acted as though they were on a sinking ship.

By the time the eighth month rolled around, the budget had nearly been expended. Gary was tired of doing everything himself. "Perhaps I should have stayed an engineer," thought Gary. Elliot Grey and Gary Anderson had a meeting to see what could be salvaged. Grey agreed to get Gary additional corporate funding to complete the project. "But performance must be met, since there is a lot riding on the Blue Spider Project," asserted Grey. He called a team meeting to identify the program status.

Anderson: "It's time to map out our strategy for the remainder of the program. Can engineering and production adhere to the schedule that I have laid out before you?"

Team member, engineering: "This is the first time that I've seen this schedule. You can't expect me to make a decision in the next 10 minutes and commit the resources of my department. We're getting a little unhappy being kept in the dark until the last minute. What happened to effective planning?"

Anderson: "We still have effective planning. We must adhere to the original schedule, or at least try to adhere to it. This revised schedule will do that."

Team member, engineering: "Look, Gary! When a project gets in trouble it is usually the functional departments that come to the rescue. But if we're kept in the dark, then how can you expect us to come to your rescue? My boss wants to know, well in advance, every decision that you're contemplating with regard to our departmental resources. Right now, we. . ."

Anderson: "Granted, we may have had a communications problem. But now we're in trouble and have to unite forces. What is your impression as to whether your department can meet the new schedule?"

Team member, engineering: "When the Blue Spider Program first got in trouble, my boss exercised his authority to make all departmental decisions regarding the program himself. I'm just a puppet. I have to check with him on everything."

Team member, production: "I'm in the same boat, Gary. You know we're not happy having to reschedule our facilities and people. We went through this once before. I also have to check with my boss before giving you an answer about the new schedule."

The following week, the verification mix was made. Testing proceeded according to the revised schedule, and it looked as though the total schedule milestones could be met, provided that specifications could be adhered to.

Because of the revised schedule, some of the testing had to be performed on holidays. Gary wasn't pleased with asking people to work on Sundays and holidays, but he had no choice, since the test matrix called for testing to be accomplished at specific times after end-of-mix.

A team meeting was called on Wednesday to resolve the problem of who would work on the holiday, which would occur on Friday, as well as staffing Saturday and Sunday. During the team meeting Gary became quite disappointed. Phil Rodgers, who had been Gary's test engineer since the project started, was assigned to a new project that the grapevine called Gable's new adventure. His replacement was a relatively new man, only eight months with the company. For an hour and a half, the team members argued about the little problems and continually avoided the major question, stating that they would first have to coordinate commitments with their bosses. It was obvious to Gary that his team members were afraid to make major decisions and therefore "ate up" a lot of time on trivial problems.

On the following day, Thursday, Gary went to see the department manager responsible for testing, in hopes that he could use Phil Rodgers this weekend.

Department manager: "I have specific instructions from the boss (director of engineering) to use Phil Rodgers on the new project. You'll have to see the boss if you want him back."

Anderson: "But we have testing that must be accomplished this weekend. Where's the new man you assigned yesterday?"

Department manager: "Nobody told me you had testing scheduled for this weekend. Half of my department is already on an extended weekend vacation, including Phil Rodgers and the new man. How come I'm always the last to know when we have a problem?"

Anderson: "The customer is flying down his best people to observe this weekend's tests. It's too late to change anything. You and I can do the testing."

Department manager: "Not on your life. I'm staying as far away as possible from the Blue Spider Project. I'll get you someone, but it won't be me. That's for sure!"

The weekend's testing went according to schedule. The raw data was made available to the customer under the stipulation that the final company position would be announced at the end of the next month, after the functional departments had a chance to analyze it.

Final testing was completed during the second week of the ninth month. The initial results looked excellent. The materials were within contract specifications, and although they were new, both Gary and Lord's management felt that there would be little difficulty in convincing the Army that this was the way to go. Henry Gable visited Gary and congratulated him on a job well done.

All that now remained was the making of four additional full-scale verification mixes in order to determine how much deviation there would be in material properties between full-sized production-run mixes. Gary tried to get the customer to concur (as part of the original trade-off analysis) that two of the four production runs could be deleted. Lord's management refused, insisting that contractual requirements must be met at the expense of the contractor.

The following week, Elliot Grey called Gary in for an emergency meeting concerning expenditures to date.

Elliot Grey: "Gary, I just received a copy of the financial planning report for last quarter in which you stated that both the cost and performance of the Blue Spider Project were 75% complete. I don't think you realize what you've done. The target profit on the program was $200,000. Your memo authorized the vice president and general manager to book 75% of that, or $150,000, for corporate profit spending for stockholders. I was planning on using all $200,000 together with the additional $300,000 I personally requested from corporate headquarters to bail you out. Now I have to go back to the vice president and general manager and tell them that we've made a mistake and that we'll need an additional $150,000."

Anderson: "Perhaps I should go with you and explain my error. Obviously, I take all responsibility."

Grey: "No, Gary. It's our error, not yours. I really don't think you want to be around the general manager when he sees red at the bottom of the page. It takes an act of God to get money back once corporate books it as profit. Perhaps you should reconsider project engineering as a career instead of program management. Your performance hasn't exactly been sparkling, you know."

Gary returned to his office quite disappointed. No matter how hard he worked, the bureaucratic red tape of project management seemed always to do him in. But late that afternoon, Gary's disposition improved. Lord Industries called to say that, after consultation with the Army, Parks Corporation would be awarded a sole-source contract for qualification and production of Spartan missile components using the new longer-life raw materials. Both Lord and the Army felt that the sole-source contract was justified, provided that continued testing showed the same results, since Parks Corporation had all of the technical experience with the new materials.

Gary received a letter of congratulations from corporate headquarters, but no additional pay increase. The grapevine said that a substantial bonus was given to the director of engineering.

During the 10th month, results were coming back from the accelerated aging tests performed on the new materials. The results indicated that although the new materials would meet specifications, the age life would probably be less than five years. These numbers came as a shock to Gary. Gary and Paul Evans had a conference to determine the best strategy to follow.

Anderson: "Well, I guess we're now in the fire instead of the frying pan. Obviously, we can't tell Lord Industries about these tests. We ran them on our own. Could the results be wrong?"

Evans: "Sure, but I doubt it. There's always margin for error when you perform accelerated aging tests on new materials. There can be reactions taking place that we know nothing about. Furthermore, the accelerated aging tests may not even correlate well with actual aging. We must form a company position on this as soon as possible."

Anderson: "I'm not going to tell anyone about this, especially Henry Gable. You and I will handle this. It will be my throat if word of this leaks out. Let's wait until we have the production contract in hand."

Evans: "That's dangerous. This has to be a company position, not a project office position. We had better let them know upstairs."

Anderson: "I can't do that. I'll take all responsibility. Are you with me on this?"

Evans: "I'll go along. I'm sure I can find employment elsewhere when we open Pandora's box. You had better tell the department managers to be quiet also."

Two weeks later, as the program was winding down into the testing for the final verification mix and final report development, Gary received an urgent phone call asking him to report immediately to Henry Gable's office.

Gable: "When this project is over, you're through. You'll never hack it as a program manager, or possibly a good project engineer. We can't run projects around here without honesty and open communications. How the hell do you expect top management to support you when you start censoring bad news to the top? I don't like surprises. I like to get the bad news from the program manager and project engineers, not secondhand from the customer. And of course, we cannot forget the cost overrun. Why didn't you take some precautionary measures?"

Anderson: "How could I when you were asking our people to do work such as accelerated aging tests that would be charged to my project and was not part of program plan? I don't think I'm totally to blame for what's happened."

Gable: "Gary, I don't think it's necessary to argue the point any further. I'm willing to give you back your old job, in engineering. I hope you didn't lose too many friends while working in program management. Finish up final testing and the program report. Then I'll reassign you."

Gary returned to his office and put his feet up on the desk. "Well," thought Gary, "perhaps I'm better off in engineering. At least I can see my wife and kids once in a while." As Gary began writing the final report, the phone rang:

Functional manager: "Hello, Gary. I just thought I'd call to find out what charge number you want us to use for experimenting with this new procedure to determine accelerated age life."

Anderson: "Don't call me! Call Gable. After all, the Blue Spider Project is his baby."

QUESTIONS

1. If you were Gary Anderson, would you have accepted this position after the director stated that this project would be his baby all the way?
2. Do engineers with MBA degrees aspire to high positions in management?
3. Was Gary qualified to be a program manager?
4. What are the moral and ethical issues facing Gary?
5. What authority does Gary Anderson have and to whom does he report?
6. Is it true when you enter project management, you either go up the organization or out the door?
7. Is it possible for an executive to take too much of an interest in an R&D project?
8. Should Paul Evans have been permitted to report information to Gable before reporting it to the program manager?
9. Is it customary for the project manager to prepare all of the handouts for a customer interchange meeting?
10. What happens when a situation of mistrust occurs between the customer and the contractor?
11. Should functional employees of the customer and contractor be permitted to communicate with one another without going through the project office?
12. Did Gary demonstrate effective time management?
13. Did Gary understand production operations?
14. Are functional employees authorized to make project decisions?
15. On R&D projects, should profits be booked periodically or at project termination?

16. Should a project manager ever censor bad news?
17. Could the abovementioned problems have been resolved if there had been a singular methodology for project management in place?
18. Can a singular methodology for project management specify morality and ethics in dealing with customers? If so, how do we then handle situations where the project manager violates protocol?
19. Could the lessons learned on success and failure during program/project debriefings cause a major change in the program/project management methodology?

The So What

- Think about the learnings form this case for how you would shape your role as a Program Manager
- Compare in your reflection the leadership style and the role of the PMO as was described in the book and how this PMO operated

CASE STUDY: MCROY AEROSPACE

McRoy Aerospace was a highly profitable company building cargo planes and refueling tankers for the armed forces. It had been doing this for more than 50 years and was highly successful. But because of a downturn in the government's spending on these types of planes, McRoy decided to enter the commercial aviation aircraft business, specifically wide-body planes that would seat up to 400 passengers, and compete head on with Boeing and Airbus Industries.

During the design phase, McRoy found that the majority of the commercial airlines would consider purchasing its plane provided that the costs were lower than the other aircraft manufacturers. While the actual purchase price of the plane was a consideration for the buyers, the greater interest was in the life-cycle cost of maintaining the operational readiness of the aircraft, specifically the maintenance costs.

Operations and support costs were a considerable expense and maintenance requirements were regulated by the government for safety reasons. The airlines make money when the planes are in the air rather than sitting in a maintenance hangar. Each maintenance depot maintained an inventory of spare parts so that, if a part did not function properly, the part could be removed and replaced with a new part. The damaged part would be sent to the manufacturer for repairs or replacement. Inventory costs could be significant but were considered a necessary expense to keep the planes flying.

One of the issues facing McRoy was the mechanisms for the eight doors on the aircraft. Each pair of doors had their own mechanisms which appeared to be restricted by their location in the plane. If McRoy could come up with a single design mechanism for all four pairs of doors, it would significantly lower the inventory costs for the airlines as well as the necessity to train mechanics on one set of mechanisms rather than four. On the cargo planes and refueling tankers, each pair of doors had a unique mechanism. For commercial aircrafts, finding one design for all doors would be challenging.

Mark Wilson, one of the department managers at McRoy's design center, assigned Jack, the best person he could think of to work on this extremely challenging program. If anyone could accomplish it, it was Jack. If Jack could not do it, Mark sincerely believed it could not be done.

The successful completion of this program would be seen as a value-added opportunity for McRoy's customers and could make a tremendous difference from a cost and efficiency standpoint. McRoy would be seen as an industry leader in life-cycle costing, and this could make the difference in getting buyers to purchase commercial planes from McRoy Aerospace.

The project was to design an opening/closing mechanism that was the same for all of the doors. Until now, each door could have a different set of open/close mechanisms, which made the design, manufacturing, maintenance, and installation processes more complex, cumbersome, and costly.

Without a doubt, Jack was the best – and probably the only – person to make this happen even though the equipment engineers and designers all agreed that it could not be done. Mark put all of his cards on the table when he presented the challenge to Jack. He told him wholeheartedly that his only hope was for Jack to take on this project and explore it from every possible, out-of-the-box angle he could think of. But Jack said right off the bat that this may not be possible. Mark was not happy hearing Jack say this right away, but he knew Jack would do his best.

Jack spent two months looking at the problem and simply could not come up with the solution needed. Jack decided to inform Mark that a solution was not possible. Both Jack and Mark were disappointed that a solution could not be found.

"I know you're the best, Jack," stated Mark. "I can't imagine anyone else even coming close to solving this critical problem. I know you put forth your best effort and the problem was just too much of a challenge. Thanks for trying. But if I had to choose one of your co-workers to take another look at this project, who might have even half a chance of making it happen? Who would you suggest? I just want to make sure that we have left no stone unturned," he said rather glumly. Mark's words caught Jack by surprise. Jack thought for a moment and you could practically see the wheels turning in his mind. Was Jack thinking about who could take this project on and waste more time trying to find a solution? No, Jack's wheels were turning on the subject of the challenging problem itself. A glimmer of an idea whisked through his brain and he said, "Can you give me a few days to think about some things, Mark?" he asked pensively.

Mark had to keep the little glimmer of a smile from erupting full force on his face. "Sure, Jack," he said. "Like I said before, if anyone can do it, it's you. Take all the time you need."

A few weeks later, the problem was solved and Jack's reputation rose to even higher heights than before.

QUESTIONS

1. Was Mark correct in what he said to get Jack to continue investigating the problem?
2. Should Mark just have given up on the idea rather than what he said to Jack?
3. Should Mark have assigned this to someone else rather than giving Jack a second chance, and if so, how might Jack react?
4. What should Mark have done if Jack still was not able to resolve the problem?
5. Would it make sense for Mark to assign this problem to someone else now, after Jack could not solve the problem the second time around?
6. What other options, if any, were now available to Mark?

The So What

- Think about the case and how one could expand the qualities of learning across other program managers in the PMO
- Consider any of the learning points and how would you coach your program sponsor

CASE STUDY: THE TEAM MEETING

Every project team has team meetings. The hard part is deciding when during the day to have the team meeting.

Know Your Energy Cycle

Vince had been a "morning person" ever since graduating from college. He enjoyed getting up early. He knew his own energy cycle and the fact that he was obviously more productive in the morning than in the afternoon.

Vince would come into work at 6:00 a.m., two hours before the normal work force would show up. Between 6:00 a.m. and noon, Vince would keep his office door closed and often would not answer the phone. This prevented people from robbing Vince of his most productive time. Vince considered time robbers such as unnecessary phone calls lethal to the success of the project. This gave Vince six hours of productive time each day to do the necessary project work. After lunch, Vince would open his office door and anyone could then talk with him.

A Tough Decision

Vince's energy cycle worked well, at least for Vince. But Vince had just become the project manager on a large project. Vince knew that he may have to sacrifice some of his precious morning time for team meetings. It was customary for each project team to have a weekly team meeting, and most project team meetings seemed to be held in the morning.

Initially, Vince decided to go against tradition and hold team meetings between 2:00 and 3:00 p.m. This would allow Vince to keep his precious morning time for his own productive work. Vince was somewhat disturbed when there was very little discussion on some of the critical issues and it appeared that people were looking at their watches. Finally, Vince understood the problem. A large portion of Vince's team members were manufacturing personnel that started work as early as 5:00 a.m. The manufacturing personnel were ready to go home at 2:00 p.m. and were tired.

The following week Vince changed the team meeting time to 11:00–12:00 noon. It was evident to Vince that he had to sacrifice some of his morning time. But once again, during the team meetings, there really wasn't very much discussion about some of the critical issues on the project and the manufacturing personnel were looking at their watches. Vince was disappointed and, as he exited the conference room, one of the manufacturing personnel commented to Vince, "Don't you know that the manufacturing people usually go to lunch around 11:00 a.m.?"

Vince came up with a plan for the next team meeting. He sent out e-mails to all of the team members stating that the team meeting would be at 11:00 a.m. to 12:00 noon as before, but the project would pick up the cost for providing lunch in the form of pizzas and salads. Much to Vince's surprise, this worked well. The atmosphere in the team meeting improved significantly. There were meaningful discussions and decisions were being made instead of creating action items for future team meetings. It suddenly became an informal rather than a formal team meeting. While Vince's project could certainly incur the cost of pizzas, salads, and soft drinks for team meetings, this might set a bad precedent if this would happen at each team meeting. At the next team meeting, the team decided that it would be nice if this could happen once or twice a month. For the other team meetings, it was decided to leave the time for the team meetings the same at 11:00 a.m. to 12:00 noon, but they would be "brown bag" team meetings where the team members would bring their lunches and the project would provide only the soft drinks and perhaps some cookies or brownies.

QUESTIONS

1. How should a project manager determine when (i.e. time of day) to hold a team meeting? What factors should be considered?
2. What mistakes did Vince make initially?
3. If you were an executive in this company, would you allow Vince to continue doing this?

The So What

- Think about how meetings' environment, timing, and setup would contribute to the motivation of a program team

CASE STUDY: THE PRIMA DONNA

Ben was placed in charge of a one-year project. Several of the work packages had to be accomplished by the Mechanical Engineering Department and required three people to be assigned full time for the duration of the project. When the project was originally proposed, the Mechanical Engineering Department manager estimated that he would assign three of his grade 7 employees to do the job. Unfortunately, the start date of the project was delayed by three months, and the department manager was forced to assign the resources he planned to use on another project. The resources that would be available for Ben's project at the new starting date were two grade 6's and a grade 9.

The department manager assured Ben that these three employees could adequately perform the required work and that Ben would have these three employees full time for the duration of the project. Furthermore, if any problems occurred, the department manager made it clear to Ben that he personally would get involved to make sure that the work packages and deliverables were completed correctly.

Ben did not know any of the three employees personally. But since a grade 9 was considered as a senior subject matter expert pay grade, Ben made the grade 9 the lead engineer representing his department on Ben's project. It was common practice for the senior-most person assigned from each department to act as the lead and even as an assistant project manager. The lead was often allowed to interface with the customers at information exchange meetings.

By the end of the first month of the project, work was progressing as planned. Although most of the team seemed happy to be assigned to the project and team morale was high, the two grade 6 team members in the Mechanical Engineering Department were disenchanted with the project. Ben interviewed the two grade 6 employees to see why they were somewhat unhappy. One of the two employees stated:

> The grade 9 wants to do everything himself. He simply does not trust us. Every time we use certain equations to come up with a solution, he must review everything we did in microscopic detail. He has to approve everything. The only time he does not micromanage us is when we have to make copies of reports. We do not feel that we are part of the team.

Ben was unsure how to handle the situation. Resources are assigned by the department managers and usually cannot be removed from a project without the permission of the department managers. Ben met with the Mechanical Engineering Department manager, who stated:

> The grade 9 that I assigned is probably the best worker in my department. Unfortunately, he's a prima donna. He trusts nobody else's numbers or equations other than his own. Whenever coworkers perform work, he feels obligated to review everything that they have done. Whenever possible, I try to assign him to one-person activities so that he will not have to interface with anyone. But I have no other one-person assignments right now, which is why I assigned him to your project. I was hoping he would change his ways and work as a real team member with the two grade 6 workers, but I guess not. Don't worry about it. The work will get done, and get done right. We'll just have to allow the two grade 6 employees to be unhappy for a little while.

Ben understood what the department manager said but was not happy about the situation. Forcing the grade 9 to be removed could result in the assignment of someone with lesser capabilities, and this could impact the quality of the deliverables from the Mechanical Engineering Department. Leaving the grade 9 in place for the duration of the project will alienate the two grade 6 employees and their frustration and morale issues could infect other team members.

QUESTIONS

1. What options are available to Ben?
2. Is there a risk in leaving the situation as is?
3. Is there a risk in removing the grade 9?

CASE STUDY: ZANE CORPORATION

Zane Corporation was a medium-sized company with multiple product lines. More than 20 years ago, Zane implemented project management to be used in all their product lines, but mainly for operational or traditional projects rather than strategic or innovation projects. Recognizing that a methodology would be needed, Zane made the faulty conclusion that a single methodology would be needed and that a one-size-fits-all mentality would satisfy almost all their projects. Senior management believed that this would standardize status reporting and make it easy for senior management to recognize the true performance. This approach worked well in many other companies that Zane knew about, but it was applied to primarily traditional or operational projects.

As the one-size-fits-all approach became common practice, Zane began capturing lessons learned and best practices with the intent of improving the singular methodology. Project management was still being viewed as an approach for projects that were reasonably well-defined, having risks that could be easily managed, and executed by a rather rigid methodology that had limited flexibility. Executives believed that project management standardization was a necessity for effective corporate governance.

THE PROJECT MANAGEMENT LANDSCAPE CHANGES

Zane recognized the benefits of using project management from their own successes, the capturing of lessons learned and best practices, and published research data. Furthermore, Zane was now convinced that almost all activities within the firm could be regarded as projects and they were therefore managing their business by projects.

As the one-size-fits-all methodology began to be applied to nontraditional or strategic projects, the weaknesses in the singular methodology became apparent. Strategic projects, especially those that involved innovation, were not always completely definable at project initiation, the scope of work could change frequently during project execution, governance now appeared in the form of committee governance with significantly more involvement by the customer or business owner, and a different form of project leadership was required on some projects. Recognizing the true status of some of the nontraditional projects was becoming difficult.

The traditional risk management approach used on operational projects appeared to be insufficient for strategic projects. As an example, strategic projects require a risk management approach that emphasizes VUCA analyses:

- **V**olatility
- **U**ncertainty
- **C**omplexity
- **A**mbiguity

Significantly more risks were appearing on strategic projects where the requirements could change rapidly to satisfy turbulent business needs. This became quite apparent on IT projects that focused heavily upon the traditional waterfall methodology that offered little flexibility. The introduction of an agile methodology solved some of the IT problems but created others. Agile was a flexible methodology or framework that focused heavily upon better risk management activities but required a great deal of collaboration. Every methodology or framework comes with advantages and disadvantages.

The introduction of an agile methodology gave Zane a choice between a rigid one-size-fits-all approach or a very flexible agile framework. Unfortunately, not all projects were perfect fits for an extremely rigid or flexible approach. Some projects were middle-of-the-road projects that fell in between rigid waterfall approaches and flexible agile frameworks.

Understanding Methodologies

Zane's original belief was that a methodology functioned as a set of principles that a company can tailor and then apply to a specific situation or group of activities that have some degree of commonality. In a project environment, these principles might appear as a list of things to do and show up as forms, guidelines, templates, and checklists. The principles may be structured to correspond to specific project life cycle phases.

For most companies, including Zane, the project management methodology, often referred to as the waterfall approach where everything is done sequentially, became the primary tool for the "command and control" of projects providing some degree of *standardization* in the execution of the work and *control* over the decision-making process. Standardization and control came at a price and provided some degree of limitation as to when the methodology could be used effectively. Typical limitations that Zane discovered included:

- *Type of project*: Most methodologies assumed that the requirements of the project were reasonably well-defined at the onset of the project. Trade-offs were primarily on time and cost rather than scope. This limited the use of the methodology to traditional or operational projects that were reasonably well-understood at the project approval stage and had a limited number of unknowns. Strategic projects, such as those involving innovation and had to be aligned to strategic business objective rather than a clear statement of work, could not be easily managed using the waterfall methodology because of the large number of unknowns and the fact that they could change frequently.
- *Performance tracking*: With reasonable knowledge about the project's requirements, performance tracking was accomplished mainly using the triple constraints of time, cost, and scope. Nontraditional or strategic projects had significantly more constraints that required monitoring and therefore used other tracking systems than the project management methodology. Simply stated, the traditional methodology had limited flexibility when applied to projects that were not operational.
- *Risk management*: Risk management was important on all types of projects. But on nontraditional or strategic projects, with the high number of unknowns that can change frequently over the life of the project, standard risk management practices that are included in traditional methodologies may be insufficient for risk assessment and mitigation practices.
- *Governance*: For traditional projects, governance was provided by a single person acting as the sponsor for the project. The methodology became the sponsor's primary vehicle for command and control and used with the mistaken belief that all decisions could be made by monitoring just the time, cost, and scope constraints.

Selecting The Right Framework

Zane recognized that the future was not simply a decision between waterfall, agile, and Scrum as to which one will be a best fit for a given project. New frameworks, perhaps a hybrid methodology, needed to be created from the best features of each approach and then applied to a project. Zane now believed with a reasonable degree of confidence that new frameworks, with a great deal of flexibility and the ability to be customized, will certainly appear in the future and would be a necessity for continued growth. Deciding which framework is best suited to a given project will be the challenge and project teams will be given the choice of which one to use.

Zane believed that project teams of the future will begin each project by determining which approach will best suit their needs. This would be accomplished with checklists and questions that address characteristics of the project such as flexibility of the requirements, flexibility in the constraints, type of leadership needed, team skill levels needed, and the culture of the organization. The answers to the questions would then be pieced together to form a framework which may be unique to a given project.

QUESTIONS

1. What are some of the questions that Zane should ask themselves when selecting a flexible methodology?
2. What issues could arise that would need resolution?
3. What would you recommend as the first issue that needs to be addressed?
4. Was it a mistake or a correct decision not allowing the sales force to manage the innovation projects?
5. Is it feasible to set up a project management methodology for managing innovation projects?

CASE STUDY: THE POOR TEAM PERFORMER

Paula, the program manager, was reasonably happy the way that work was progressing on the program. The only issue was the work being done by Frank. Paula knew from the start of the program that Frank was a mediocre employee and often regarded as a trouble-maker. The tasks that Frank was expected to perform were not overly complex and the line manager assured Paula during the staffing function that Frank could do the job. The line manager also informed Paula that Frank demonstrated behavioral issues on other programs and sometimes had to be removed from the program. Frank was a chronic complainer and found fault with everything and everybody. But the line manager also assured Paula that Frank's attitude was changing and that the line manager would get actively involved if any of these issues began to surface on Paula's program. Reluctantly, Paula agreed to allow Frank to be assigned to her program.

Unfortunately, Frank's work on the program was not being performed according to Paula's standards. Paula had told Frank on more than one occasion what she expected from him, but Frank persisted on doing his own thing. Paula was now convinced that the situation was getting worse. Frank's work packages were coming in late and sometimes over budget. Frank continuously criticized Paula's performance as a program manager and Frank's attitude was beginning to affect the performance of some of the other team members. Frank was lowering the morale of the team. It was obvious that Paula had to take some action.

QUESTIONS

1. What options are available to Paula?
2. If Paula decides to try to handle the situation first by herself rather than approach the line manager, what should Paula do and in what order?
3. If all of Paula's attempts fail to change the worker's attitude and the line refuses to remove the worker, what options are available to Paula?
4. What rights, if any, does Paula have with regard to wage and salary administration regarding this employee?

The So What

- Think about your role as a program manager and the learning from this simple case that could affect your program team building activities and associated communications excellence

CASE STUDY: THE MANAGEMENT CONTROL FREAK

The company hired a new vice president for the Engineering Department, Richard Cramer. Unlike his predecessor, Richard ruled with an iron hand and was a true micromanager. This played havoc with the project managers in Engineering because Richard wanted to be involved in all decisions, regardless of how small.

What To Do

Anne was an experienced project manager who had been with the company for more than 20 years. She had a reputation for being an excellent project manager and people wanted to work on her projects. She knew how to get the most out of her team and delegated as much decision-making as possible to her team members. Her people skills were second to none.

A few months before Richard Cramer was hired Anne was assigned to a two-year project for one of the company's most important clients. Anne had worked on projects for this client previously, and the results were well received by the client. The client actually requested that Anne be assigned to this project.

Almost all of Anne's team members had worked for her before. Some of the team members had even asked to work for her on this project. Anne knew some of the people personally and trusted their decision-making skills. Having people assigned that have worked with you previously is certainly considered a plus.

Work progressed smoothly until about the third week after Richard Cramer came on board. In a meeting with Anne, Richard commented:

I have established a policy that I will be the project sponsor for all projects where the project managers report to someone in Engineering. I know that the Vice President for Marketing had been your sponsor for previous projects with this client, but all of that will now change. I have talked with the Vice President for Marketing and he understands that I will now be your sponsor. I just cannot allow anyone from outside of Engineering to be a sponsor of a project that involves critical engineering decisions and where the project managers come from Engineering. So, Anne, I will be your sponsor from now on and I want you to talk to my secretary and set up weekly briefings for me on the status of your project. This is how I did it in my previous company and it worked quite well.

These comments didn't please Anne. The vice president for marketing was quite friendly with the client and now things were changing. Anne understood Richard's reasons for wanting to do this but certainly was not happy about it.

Over the next month, Anne found that her working relationship with Richard was getting progressively worse, and it was taking its toll on the project. Richard was usurping Anne's authority and decision-making. On previous projects, Anne would meet with the sponsor about every two weeks and the meeting would last about 15 minutes. Her meetings with Richard were now weekly and were lasting for more than one hour. Richard wanted to see all of the detailed schedules and wanted a signature block for himself on all documents that involved engineering decisions. There was no question in Anne's mind that Richard was a true micromanager.

At the next full team meeting, some of the workers were complaining that Richard was calling them directly, without going through Anne, and making some decisions that Anne did not know about. The workers were receiving directions from Richard that were in conflict with directions provided by Anne. Anne could tell that morale was low and heard people mumbling about wanting to get off of this project.

At Anne's next meeting with Richard, she made it quite clear about how upset she was with Richard's micromanagement of the project and, if this continued, she would have a very unhappy client. Richard again asserted how he had to be involved in all technical decisions and that this was his way of managing. He also stated that if Anne was unhappy, he could find someone else to take over her job as the project manager.

Something had to be done. This situation could not be allowed to continue without damaging the project further. Anne thought about taking her concerns directly to the president but realized that nothing would probably change. And if that happened, Anne could be worse off.

Anne then came up with a plan. She would allow Richard to micromanage and even help him do so. There was a risk in doing this and Anne could very well lose her job. But she decided to go ahead with her plan. For the next several weeks, Anne and all of the team members refused to make even the smallest decisions themselves. Instead, they brought all of the decisions directly to Richard. Richard was even getting phone calls at home from the team members on weekends, during the dinner hour, late at night, and early Sunday mornings.

Richard was now being swamped with information overload and was spending a large portion of his time making mundane decisions on Anne's project. In the next team sponsor briefing meeting with Anne, Richard stated:

> I guess that you've taught me a lesson. "If it's not broken, then there isn't any reason to fix it." I guess that I came across too strong and made things worse. What can we do to repair the damage I may have done?

Anne could not believe that these words were spoken by Richard. Anne was speechless. She thought for a moment and then went over to the white board in Richard's office. She took a magic marker and drew a vertical line down the center of the board. She put her name to the left of the line and Richard's name to the right of the line. She then said:

> I'm putting my responsibilities as a project manager under my name and I'd like you to put your responsibilities as a sponsor under your name. However, the same responsibility cannot appear under both names.

An hour later, Anne and Richard came to an agreement on what each other's responsibilities should be. Anne walked out of Richard's office somewhat relieved that she was still employed.

QUESTIONS

1. When someone hires into a company, is there any way of telling whether or not they are a control freak?
2. If someone higher in rank than you turns out to be a control freak, how long should you wait before confronting them?
3. Do you believe that Anne handled the situation correctly?
4. Could Anne's decision on how to handle the situation result in Anne getting removed as a project manager or even fired?
5. What other ways were available to Anne for handling the situation?

The So What

- This case should have been an excellent example of educating upward
- Anne has showcased how to work with your project/program sponsor
- Establishing this critical partnership on a strong foundation as discussed in the book, is one of you most-valuable success enablers

As a closing exercise and as you review the previous eight case studies, try to breakdown your learnings into the four buckets representing the major four sections of the book. There should be takeaways pertaining to driving strategic change, the Power Skills, the PMO impact creating role, and linkages between strategy execution to expected value and joint view of success. The topics and skills addressed in this book require a conscientious decision on your part as a leader to strengthen the muscles of reflecting, quick applications, and making fit decisions for the future program/project scenarios you encounter.

LEADING AND SUSTAINING FUTURE CHANGE

As seen throughout the chapters of this book, leading and sustaining future changes requires some critical ingredients for that future of work. Delivering change and transformative execution outcomes, will have to build on a strong foundation of the most fitting culture for the mission of the organization and its stakeholders. Value achievement is at the center of these future cultures. Sometimes, the value of a program is not determined until well into commercialization and that's where value is experienced. The ***Program Way*** with the balanced trio of people, process, and technology anchors of the future, will have a direct impact on how we work, where we spend our time, and what dealing with the elevated value expectations looks like.

Project and Program Management Cultures

Introduction to Cultures

All companies have organizational or corporate cultures, but the definition of a culture can change based upon who is doing the investigation. Organizational cultures are usually described in terms of the vision, values, beliefs, and norms that senior managers exhibit as to how they wish to exercise command and control from the top floor of the building now and possibly in the future. The culture describes the behavior management expects people to follow when interfacing with other employees, customers, and stakeholders.

Every company has its own unique culture. However, in large organizations, there can exist conflicting subcultures created by different management teams and designed to satisfy the needs of certain functional organizations or business objectives. Based upon the leadership styles, the subcultures may be in conflict. Cultures for projects and programs are often treated in the literature as subcultures to the organizational culture.

Some of the factors that make up a culture are often attributed to history, past performance, type of products and services, industry expectations, customer base, technology, strategic planning activities, type of employees and their knowledge level, management leadership styles, and possibly a need for coexistence within a national culture.[3]

Benefits of Cultures

The benefits and perception of an effective culture can be different based upon whose eyes are looking at it. Creating the ideal culture is difficult, if not impossible. There will always be people that believe the culture can be improved. Almost every type of culture will have some benefits.

Executives may view strong cultures as providing the following benefits:[4]

- Better aligning the company toward achieving its vision, mission, and goals
- Can lead to a competitive advantage
- High employee motivation and loyalty
- Increased team cohesiveness among the company's various departments and divisions
- Promoting consistency and encouraging coordination and control within the company
- Shaping employee behavior at work, enabling the organization to be more efficient

Workers might see an effective culture as:

- A way to share knowledge across the entire organization
- A chance to demonstrate one's leadership potential
- The opportunity for long-term employment
- Ability to share one's ideas with colleagues
- Ability to participate in decisions
- The chance to achieve one's desires and aspirations

Designing the Culture

Establishing an organizational or corporate culture takes time especially if people are removed from their comfort zones and are expected to act in manner they do not prefer. The word "time" can be a blessing or a curse when trying to create a culture for the organization, projects, or programs. Even though senior managers have their own ideas as to what the culture should be, it takes time to create the culture and get buy-in from the work force.

Buy-in requires showing the work force that they will benefit from following and adapting to the organizational culture which can help them cope with internal and external problems. When buy-in is successful, workers can more easily identify with the organization and how they must interact. Successful organizational cultures represent a "corporate personality" (Flamholtz and Randle 2011).

Corporate or organizational cultures usually take the longest to design and implement. Prior to a cultural change initiative, a needs assessment is needed to identify and understand the current organizational culture. This can be done through employee surveys, interviews, focus groups, observation, customer surveys where appropriate, and other internal research, to further identify areas that require change. The company must then assess and clearly identify the new, desired culture, and then design a change process.[5]

[3] Adapted from Wikipedia: Organizational Culture

[4] Ibid

[5] Adapted from Wikipedia: Organizational Culture

Organizational cultures are influenced heavily by the firm's business model. It may take several months or longer to design the correct culture and even then, there is no guarantee that the workers will readily accept the changes needed.

When we talk about organizational cultures, we normal assume we are discussing one culture for the entire organization, and the culture will last for an indefinite amount of time. But when discussing project and program organizational cultures, the life cycle time factor becomes important.

Projects have a finite life cycle. As such, project managers may not have sufficient time to develop the ideal project culture. At the kickoff meeting for the project, the team members may be told by the project manager what the culture will be. The workers may not like the culture that the project manager expects, but the workers know that the project will eventually come to an end. The result is that they usually endure the pain, if any, and look forward to their next assignment.

Programs usually have a longer time duration than projects and are more closely aligned to an organizational culture. The program manager may have sufficient time to perform an assessment and design a program culture that the program team members will readily accept.

Some cultures may take a long time to create and put into place but can be torn down overnight. Cultures for project and program management are based more so on organizational behavior, not processes. Corporate cultures reflect the goals, beliefs, and aspirations of senior management. It may take years for the building blocks to be in place for a good culture to exist, but that culture can be torn down quickly through the personal whims of one executive who refuses to support project/program management practices.

Project/program management cultures can exist within any organizational structure. The speed at which the culture matures, however, may be based on the size of the company, the size and nature of the projects, and the type of customer, whether the customer is internal or external. Project and program management practices are a culture, not policies and procedures. As a result, it may not be possible to benchmark project and program management cultures. What works well in one company may not work equally well in another.

Strong cultures can form when project and program management practices are viewed as a profession and supported by senior management. A strong culture can also be viewed as a primary business differentiator. Strong cultures can focus on either a formal or an informal project/program management approach. However, with the formation of any culture, there are always some barriers that must be overcome.

Within excellent companies, the process of project and program management have evolved into a behavioral culture based on multiple-boss reporting. Some of the early research in project and program management cultures focused heavily on the culture needed to make matrix management practices work effectively (Elmes and Wilemon 1988; Morrison et al. 2006; Cleland 1999).

Today, there is significantly more interest in project and program management cultures than in the past. Organizations are now working on significantly more projects than in the past. Most of the projects and programs now are composed of strategic, R&D or innovation projects that have much longer time durations than in the past. The result is a greater need for effective cultures with longer life cycles.

Types of Organizational Cultures

There are many types of organizational cultures based upon the methods used to classify the organizational culture. Below are a few of the many types of organizational cultures that appear in the literature:

- *Strong cultures:* These are clearly understood by everyone. The workers use the organization's tools and processes, often without question. Usually, this is accompanied by high motivation and loyalty.
- *Weak cultures:* These cultures must be enforced by policies, procedures, and bureaucracy because there may be little alignment to corporate values.
- *Adaptive cultures:* These cultures can change quickly when necessary.
- *Nonadaptive cultures:* These cultures are risk adverse and usually do not take advantage of opportunities.
- *Bullying cultures:* These cultures may allow for abusive behavior to exist. This culture must be supported by senior management
- *National cultures:* These cultures are deep rooted values how companies should be run.

While there is no single "type" of organizational culture and organizational cultures vary widely from one organization to the next, commonalities exist, and some researchers have developed models to describe different desired indicators of organizational cultures. Organizations should strive for what is considered a "healthy" organizational culture to increase productivity, growth, efficiency, and reduce counterproductive behavior and turnover of employees. A variety of characteristics describing a healthy culture include:[6]

- Acceptance and appreciation for diversity
- Regard for fair treatment of each employee as well as respect for each employee's contribution to the company
- Employee pride and enthusiasm for the organization and the work performed
- Equal opportunity for each employee to realize their full potential within the company
- Strong communication with all employees regarding policies and company issues
- Strong company leaders with a strong sense of direction and purpose
- Ability to compete in industry innovation and customer service, as well as price
- Lower than average turnover rates (perpetuated by a healthy culture)

Types of Project and Program Management Cultures

Perhaps the most significant characteristic of companies that are excellent in project management is their culture. Successful implementation of project management creates an organization and cultures that can change rapidly because of the demands of each project and yet adapt quickly to a constantly changing dynamic environment, perhaps at the same time. Successful companies must cope with change in real time and live with the potential disorder that comes with it. The situation can become more difficult if two companies with possibly diverse cultures must work together on a common project.

Change is inevitable in all organizations but perhaps more so in project-driven organizations. As such, excellent companies have come to the realization that competitive success can be achieved only if the organization has achieved a culture that promotes and sustains the necessary organizational behavior. Corporate cultures cannot be changed overnight. The time frame is normally years but can be reduced if executive support exists. Also, if as little as one executive refuses, to support a potentially good project management culture, disaster can result.

There are different types of project/program management cultures, which vary according to the nature of the business, the amount of trust and cooperation, and the competitive environment. Some typical types of cultures include

- *Cooperative cultures.* These are based on trust and effective communication, not only internally but externally as well with stakeholders and clients.
- *Noncooperative cultures.* In these cultures, mistrust prevails. Employees worry more about themselves and their personal interests than what is best for the team, company, or customer.
- *Competitive cultures.* These cultures force project/program teams to compete with one another for valuable corporate resources. In these cultures, project/program managers often demand that employees demonstrate more loyalty to the project than to their line manager. This can be disastrous when employees are working on multiple projects at the same time and receive different instructions from the project and the functional manager.
- *Isolated cultures.* These occur when a large organization allows functional units to develop their own project/program management cultures. This could also result in a culture-within-a-culture environment within strategic business units. It can be disastrous when multiple isolated cultures must interface with one another.
- *Fragmented cultures.* Projects or programs where part of the team is geographically separated from the rest of the team may lead to a fragmented culture. Virtual teams are often considered fragmented cultures. Fragmented cultures also occur on multinational projects, where the home office or corporate team may have a strong culture for project management, but the foreign team has no sustainable project management culture.

[6] Wikipedia: Organizational Cultures

Cooperative cultures thrive on effective communications, trust, and cooperation. Decisions are made based on the best interest of all of the stakeholders. Executive sponsorship, whether individual or committee, is more passive than active, and very few problems ever go up to the executive levels for resolution. Projects are managed more informally than formally, with minimum documentation, and often meetings are held only as needed. This type of project/program management culture takes years to achieve and functions well during both favorable and unfavorable economic conditions.

Noncooperative cultures are reflections of senior management's inability to cooperate among themselves and possibly their inability to cooperate with the workforce. Respect is nonexistent. Noncooperative cultures can produce a good deliverable for the customer if the end justifies the means. However, this culture does not generate the number of project successes achievable with the cooperative culture.

Challenges with Project/Program Management Cultures

Creating an effective culture for projects and programs requires an understanding of the challenges that must be overcome. The biggest challenge is usually in the requirements given to the project and program managers at project initiation. If the time and cost are unrealistic from the start, developing a strong or healthy culture will be difficult. Other challenges that must be overcome include

- How much creativity will be required?
- If creativity is required, will it be individual or group creativity?
- Will qualified resources be available?
- If the resources are part-time, will they be able to coexist with duties on other projects or work assignments in their functional area?
- How will the project team members react if they receive conflicting instructions from the project/program manager and their functional manager who performs their performance reviews?
- Will the team have access to all the information they need?
- Will the team members be willing to share information?
- Will team members be empowered with the authority to make certain decisions? If not, who will possess the authority?
- Will the project and program manager's authority be centralized or decentralized?
- Will the team members accept the leadership style that the project/program manager will be using?
- Will the team understand the strategic direction and support it?
- Will the team focus on short- or long-term strategic objectives when making decisions?
- How much cross-functional integration and teamwork will be required?
- How much multinational information integration will be required?
- How can we determine if team members' personal expectations from working on this project can be met, such as career path opportunities?
- Will the team have a tolerance for risk?
- How difficult will it be for team members to accept changes that might remove them from their comfort zones?

Addressing all of these challenges in a short amount of time is almost impossible. However, there is hope. While the project and program managers may be able to create a cooperative team culture, it must be supported by a similar corporate or organizational culture that encourages ideas to flow freely, understands the strengths and weaknesses of the personnel, and has confidence in their abilities. Project and program management cultures are usually subcultures to the organizational culture. As stated previously, there are some generalities among all of the cultures.

Table 1 identifies some of the generalities that senior management can implement that can make it easy for project and program managers to establish meaningful, healthy cultures is a shorter amount of time.

TABLE 1 Some High-Level Components of a Culture

Component	Description
Vision	Team members are provided with and share clear and valued objectives aligned to the business strategy
Participative Safety	A nonthreatening environment exists where team members can participate freely in discussions and decisions without fear of reprimand
Task Orientation	A focus on achieving excellence through high-quality work and acceptance of constructive criticism
Tolerance for Failure	Recognition that on some projects and programs a significant amount of risk exists and may not be able to be mitigated

Three Anchors for the Program Way

To recap a few of the critical muscle building skills, growth mindset, and leadership capabilities building in your future role as a leader and a program/project manager, the following three anchors summarize the ***Program Way*** and bring together some of the critical reminders and questions that are necessary for delivering change in tomorrow's organizations and across successful program teams.

Anchor 1: People – Customer Centricity

Figure 1 identifies, using a customer centricity focus, the key attributes, focus points, and questions we should use as we deliver products, services, or solutions. Whether the customer is internal or external, these reminders in the map shown would enable us to remember what is takes to put people and program stakeholders first. Many organizations and customers are considering what they would refer to as value-in-use. After program and project deliverables are shipped, members of the project/program teams observe how the users interface with their products and use them, and then ask these users what value they see and cherish.

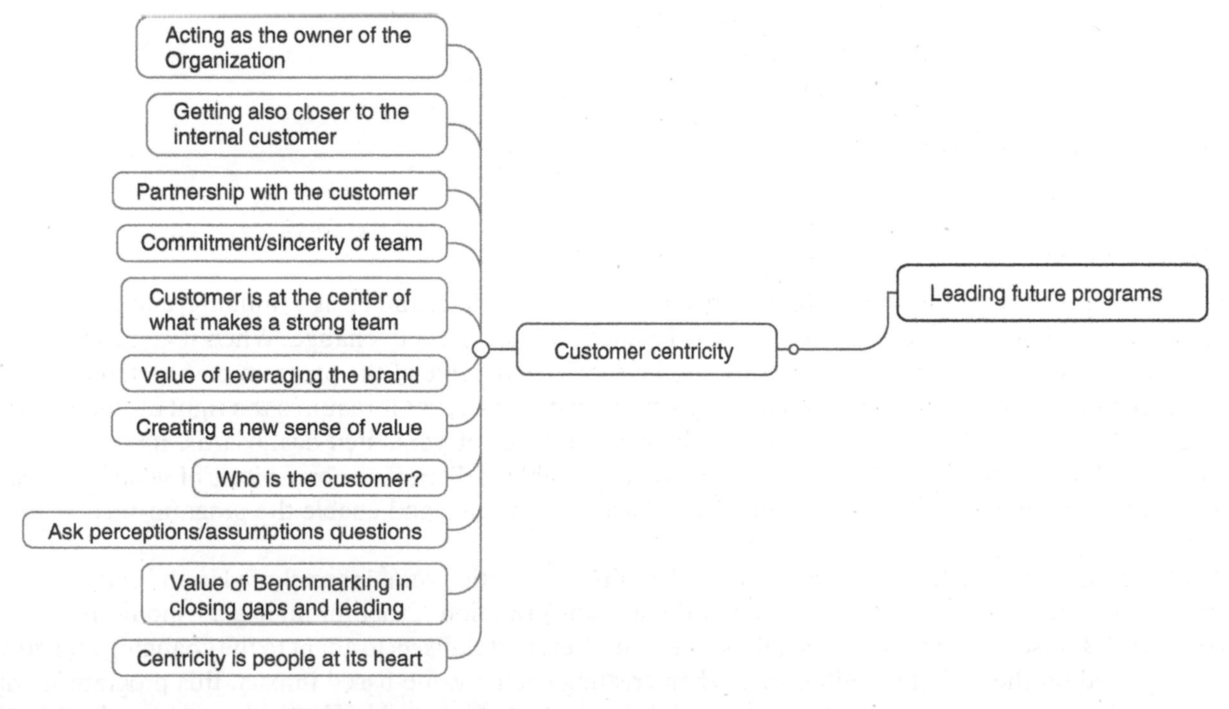

FIGURE 1 The People Anchor (With a Focus on the Customer)

Anchor 2: Agile Processes

Figure 2 breaks down leading with agility into the buckets of D-VUCAD, where VUCA is combined with disruption and diversity, indicating the kind of a future of work ahead of organizations, and the bucket of mindset. These two buckets and their associated reminders for the shifts we should consider in the way of working and how we think, should provide a comprehensive checklist to follow in preparing ourselves for the type of situational and fitting leadership of future programs.

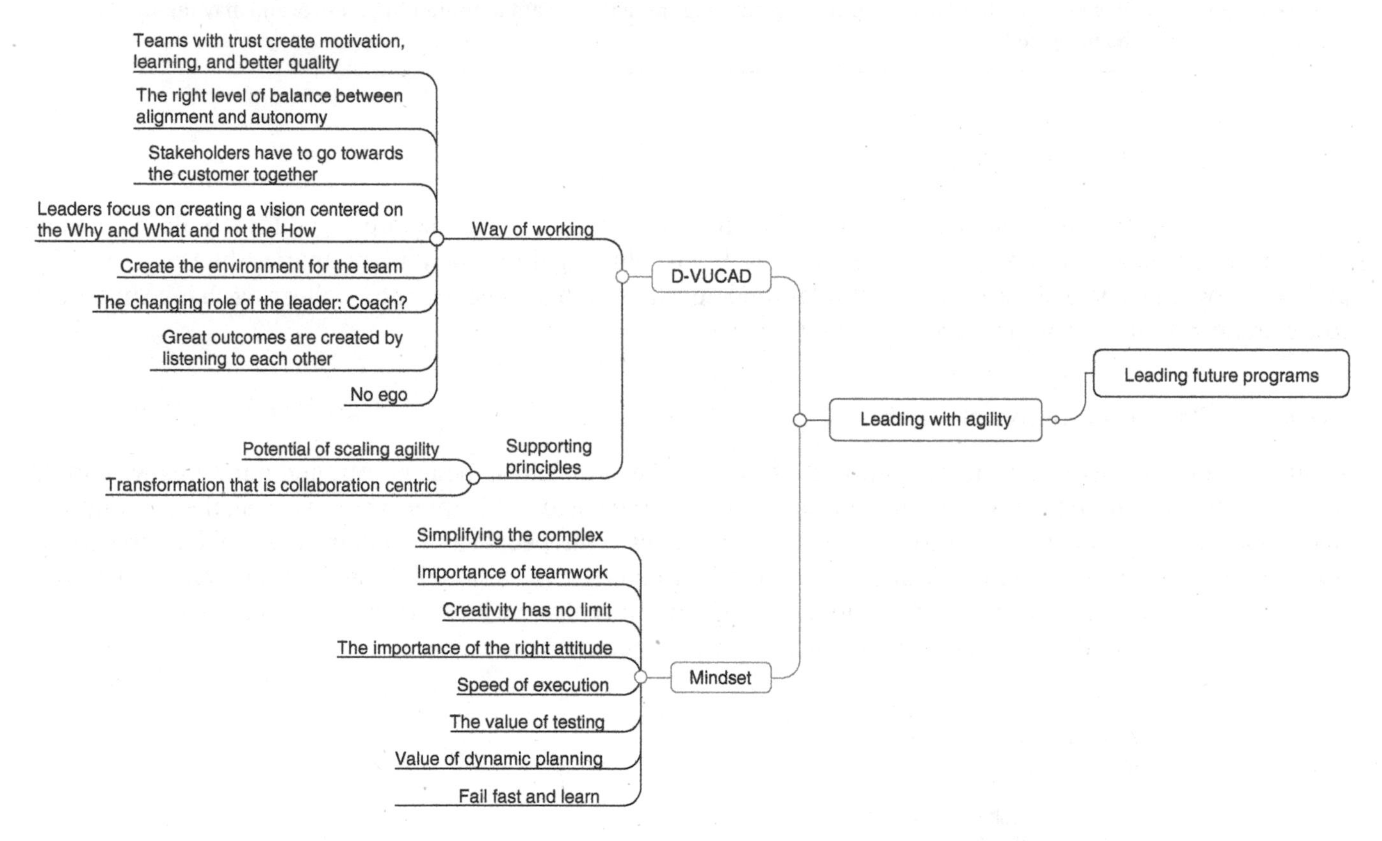

FIGURE 2 Leading with Agility

Anchor 3: Digitalization

Figure 3 highlights three groups of enablers across experimenting, data, and design thinking, covered across various points throughout the book. The key here is balance and openness to change. When it comes to Artificial Intelligence applications in program management, for example, the role of the program manager, the PMO, and the enhanced planning and execution accuracy potential is immense, yet it requires the right attitude as we say in the previous anchors. With ChatGPT, do we look at the value of generated dashboards, reports, and other governance-related assets programs today still use, or are we able to disrupt our own views of what matters. This is expected to tremendously enhance the quality of ideas and insights and enable the program success ingredients, previously discussed, to come to fruition.

The future of program management is strong. Its value proposition remains anchored in the endless need for accomplishing critical change missions and continual transformation. Program managers should remain on a path of relentless discovery and learning, adjust their mindsets and skillsets to adapt to the changing and growing demands placed on these change initiatives. When creating such a value-based impact, this program manger's role could be highly rewarding in multiple ways, while being strategically valued by the future organizations. This next generation of like-minded leaders are the torchbearers of the ***Program Way.***

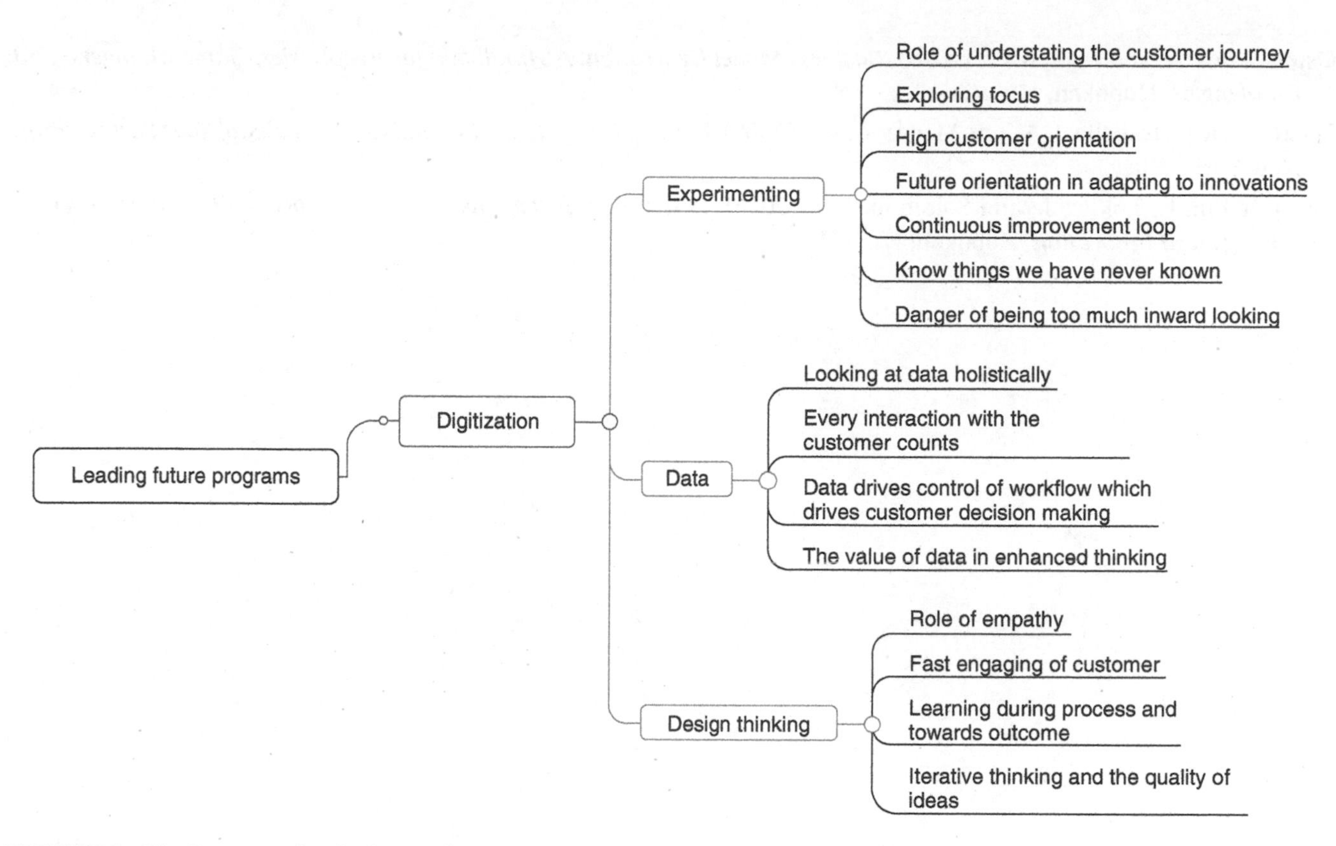

FIGURE 3 The Future Technologies Anchor

REFERENCES

Abdelkafi, N., Makhotin, S., and Posselt, T. (2013). Business Model Innovations for electric mobility – What can be learned from existing Business Model Patterns? *International Journal of Innovation Management* 17 (1): 1–41.

Baden-Fuller, C. and Morgan, M. (2010). Business models as models. *Long Range Planning* 43 (2): 156–171.

Beinhocker, E.D. (2007). *The Origin of Wealth: The Radical Remaking of Economics and What It Means for Business and Society*. Cambridge: Harvard Business School Press.

Cleland, D.I. (1999). *Project Management: Strategic Design and Implementation*, 3e. New York: McGraw Hill.

Elmes, M. and Wilemon, D. (1988). Organizational culture and project leader effectiveness. *Project Management Journal* **XIX** (1): 54–63.

Flamholtz, E.G. and Randle, Y. (2011). *Corporate Culture: The Ultimate Strategic Asset. Stanford Business Books*, 6. Stanford, CA: Stanford University Press.

Gassmann, O., Frankenberger, K., and Csik, M. (2013). *Geschäftsmodelle Entwickeln: 55 Innovative Konzepte mit dem St. Galler Business Model Navigator*. St. Gallen: Carl Hanser Verlag GmbH Co KG.

Kaplan, S. (2012). *The Business Model Innovation Factory*. Hoboken, NJ: Wiley.

Kerzner, H. (2022). *Project Management Case Studies*, 6e. Wiley.

Kerzner, H. (2023). *Innovation Project Management*. Hoboken, NJ: Wiley; Chapter 8.

Lenfle, S. (2008). Exploration and project management. *International Journal of Project Management* 26 (5): 469–478.

Lüttgens, D. and Diener, K. (2016). Business model patterns used as a tool for creating (new) innovative business models. *Journal of Business Models* 4 (3): 19–36.

Morrison, J.M., Brown, C.J., and Smit, E.v.d.M. (2006). A supportive organisational culture for project management in matrix organisations: A theoretical perspective. *South African Journal of Business Management* **37** (4): 39–54.

Osterwalder, A. and Pigneur, Y. (2010). *Business Model Generation: A Handbook for Visionaries, Game Changers, and Challengers*. Hoboken, NJ: Wiley.

Spekman, R.E., Isabella, L.A., and MacAvoy, T.C. (2000). *Alliance Competence: Maximizing the Value of Your Partnerships*. New York: Wiley.

Van Der Pilj, P., Lokitz, J., and Solomon, L.K. (2016). *Designing a Better Business: New Tools, Skills and Mindset for Strategy and Innovation*. Hoboken, NJ: Wiley.

Index

Program Management: Going Beyond Project Management to Enable Value-Driven Change, First Edition. Al Zeitoun.